3rd International Conference on Sustainable and Renewable Energy Engineering (ICSREE 2018)

E3S Web of Conferences Volume 57 (2018)

Zamora, Spain
9-11 May 2018

Editor:

Nezihe Ayas

ISBN: 978-1-5108-7295-0

Printed from e-media with permission by:

Curran Associates, Inc.
57 Morehouse Lane
Red Hook, NY 12571

Some format issues inherent in the e-media version may also appear in this print version.

Printed by Curran Associates, Inc. (2018)

For additional information, please contact EDP Sciences – Web of Conferences
at the address below.

EDP Sciences – Web of Conferences
17, Avenue du Hoggar
Parc d'Activité de Courtabœuf
BP 112
F-91944 Les Ulis Cedex A
France

Phone: +33 (0) 1 69 18 75 75
Fax: +33 (0) 1 69 28 84 91

contact-edps@webofconferences.org

Additional copies of this publication are available from:

Curran Associates, Inc.
57 Morehouse Lane
Red Hook, NY 12571 USA
Phone: 845-758-0400
Fax: 845-758-2633
Email: curran@proceedings.com
Web: www.proceedings.com

TABLE OF CONTENTS

Preface

We are pleased to present the volume of 2018 3rd International Conference on Sustainable and Renewable Energy Engineering. The conference ICSREE2018 was held in the University of Salamanca, Zamora, Spain during May 9-11, 2018; with the aim to create a linkage between industry and academia; the conference is dedicated to building an overarching technology platform for researchers in academics. As part of the event, the ICSREE issue meant to look far out enough in time, space and across disciplines and focusing on energy materials science, environmental engineering, clean energy and new energy technology, solar energy utilization technology.

The conference program was put together as a result of much hard work and dedication by many individuals. With the important efforts of the whole committee, the evaluation of all the papers was performed based on the reports from anonymous reviewers, who are qualified in the field of sustainable and renewable energy. We envision that the volume will provide an opportunity for readers to engage with a selection of refereed papers that were presented during the ICSREE2018 conference.

This conference has also been highlighted by the guest speakers: Prof. David T.W. Lin, from National University of Tainan, talked on *A Similar Capacity Test Model of Geothermal Energy*; Prof. Jesus Toribio, from University of Salamanca, Spain, talked on *Cold Drawn Pearlitic Steel Wires for Wind Turbines Structures: In the Wake of Miguel de Cervantes and Johann Sebastian Bach*; Prof. Nezihe Ayas, from Eskisehir Technical University, Turkey, talked on *Hydrogen Production From Biomass Gassification.*

Special thanks would be extended to the University of Salamanca, Spain for the support of the conference, including providing conference venue, arranging illuminating visits in the labs and campus.

Our sincere gratitude goes to the session chairs and the technical program committee members for their hard efforts in ensuring the paper quality.

We hope that you enjoy and find valuable your engagement with the ideas in sustaining your own professional development in the global world of engineering energies.

The 2019 4th International Conference on Sustainable and Renewable Energy Engineering will be held in Beijing, China. We hope that ICSREE will provide the incentive to experiment and sufficient motivation to continue making progress.

Conference Program Chair
Prof. Dr. Nezihe Ayas,
Eskisehir Technical University, Turkey

Optimization of Enhanced Geothermal System Power Generation Investment Scheme Based on Fuzzy VIKOR Model

Yan Bu[1] , *Mingjing* Guo[1], *Penghong* Cheng[2]

[1]School of Economics and Management, China University of Geosciences (Wuhan), China
[2]School of Management, Huazhong University of Science and Technology, China

Abstract. In the future, Enhanced Geothermal System (EGS) power generation will be a hot spot for new energy investment development. How to optimize the EGS power generation investment program is an important issue to be solved by the government and enterprises. In order to improve the accuracy of the EGS investment project evaluation, this paper proposes a technical route to optimize the EGS power generation investment program. It comprehensively considers the ambiguity of the decision attributes and the conflict between the indicators. The index is calculated using the entropy method and the order relationship method. Comprehensive weights; using triangular fuzzy numbers instead of exact numbers to reduce the loss of decision information; using a multi-criteria compromised sorting method to obtain trade-off solutions to solve the problem that investment plans cannot be accepted by all decision makers when there are multiple conflicting indicators. Finally, taking an enterprise's EGS power generation project as an example, the effectiveness and rationality of the optimized technology route for the EGS power generation investment program are verified.

1 Introduction

EGS (Enhanced Geothermal System), including the initial concept of Hot Dry Rock (HDR), was originated at the Los Alamos National Laboratory (LANL) in the USA [1]. The patent on the concept, which dates from 1974, describes the formation of an entirely natural tank designed for obtaining geothermal energy. In more than 40 years of research and development of the EGS, the utilization technology has matured day by day. The development prospect is very promising because of its great value. The potential of EGS resources in China is huge and the prospect of development is wide, which is higher than that estimated by the US(57 trillion tons of standard coal).Preliminary estimates indicate that EGS resources in the 3-10km range are equivalent to 860 trillion tons of standard coal, of which 2 percent is equivalent to 4,040 times China's total energy consumption in 2010 [2-3].In particular, at 3.5-7.5 km, the temperature of the EGS resources ranged from 150-250℃, and the amount of resources was huge, equivalent to 215 trillion tons of standard coal. In recent years, China has raised its green development strategy to an unprecedented level. The exploitation and utilization of geothermal resources are faced with historic development opportunities and great potential. EGS, as the clean energy with the most application value and potential for use, will play a more important role in optimizing energy structure, preventing and controlling atmospheric pollution, improving people's lives, and promoting the scientific and coordinated development of the economic and social ecological environment [4].

From the perspective of power construction costs, geothermal power generation is roughly 10,000 RMB installed 1KW, which is equivalent to the cost of wind power generation, but compared with solar photovoltaic power generation of about 100,000 RMB installed 1KW, the construction cost is much lower [5]. From the viewpoint of environmental protection value, the utilization ratio of EGS resources can reach 80%. Calculated according to the recoverable amount of 2% of China's EGS resources, it can replace the standard coal of 14.298 trillion tons, reducing CO_2 emissions by 0.2388 trillion RMB, reduce the emission of NO_x by 0.08579 trillion tons, reduce the emission of suspended dust by 0.1144 trillion tons, and reduce the coal ash residue by 1.4298 trillion tons [6]. From the perspective of natural conditions, the development and utilization of wind power, solar energy and tidal energy are all closely related to the climatic conditions, which often results in the instability and discontinuity of development and utilization. The development and utilization of geothermal energy do not have such problems.

For a long time in the future, as a significantly effective way of electricity generation, EGS will be a hot spot for new energy investment and development. In the face of a variety of EGS power generation investment plans, more comprehensive and reasonable decision-making system are needed. The optimization of EGS power generation investment plan is a multi-criteria decision-making problem. The decision-making process needs to consider many factors such as resources, economy, environment and risk. The traditional multi-criteria decision-making problems mainly include analytic hierarchy process (AHP), analytic network process (ANP) and technique for order preference by similarity to ideal solution (TOPSIS). However, there are some problems in the application of existing multi-criteria decision-making methods to the selection of EGS power generation projects, which will reduce the accuracy of decision-making. On the one hand, due to the fact that the EGS power generation industry is still commercialized in China, there are few data available for reference in the investment decision-making process, and it is difficult to obtain the precise data needed for general methods. On the other hand, due to the complexity of decision-making issues and the conflict between the indicators of the EGS power generation project, the evaluation results may not be accepted by all policy makers. Based on the above problems, this paper proposes a preferred framework for the EGS power generation investment plan based on fuzzy set theory [7-8] and multi-criteria compromised sorting method [9-11].The use of triangular fuzzy numbers to represent the decision values of each candidate item prevents the loss of decision information. Through the multi-criteria compromised sorting method, ranking of a series of solutions with multiple conflicts or inconsistent indicators can be achieved, so as to better complete the optimization of the investment scheme of the EGS power generation.

2 EGS power generation investment program evaluation index system

The construction of investment evaluation index system is the basic work for optimizing the investment project, and whether the index system is properly constructed will directly affect the accuracy of the project selection. EGS power generation project includes many aspects, especially when it is a government investment project, it needs to consider social and environmental factors. This paper will build 25 indicators through the five criteria of strengths, quality, economy, risk, and environment to evaluate a power generation project in EGS, as shown in Table 1.

Table 1. Evaluation index system of EGS power generation investment plan.

Guidelines	Primary indicators	Secondary indicators
(A) Advantage	(a) Resources	(a_1) Exploitable resources
		(a_2) EGS temperature
	(b) Location	(b_1) Geological terrain conditions
		(b_2) Mining depth
		(b_3) Grid conditions
(B) Quality	(c) EGS system	(c_1) Annual installed capacity
		(c_2) Technical maturity
		(c_3) Adaptability to local resources
		(c_4) Resistance to geological disasters
	(d) Construction	(d_1) Construction plan
		(d_2) Construction management level
(C) Economic	(e) Investment	(e_1) Fixed asset investment
		(e_2) Liquidity investment
		(e_3) Annual operating costs
	(f) Benefit	(f_1) Internal Rate of Return(%)
		(f_2) Payback period (a)
		(f_3) Policy subsidies
(D) Risk	(g) Social risk	(g_1) Policy change risk
		(g_2) Project start-up operational risk
	(h) Technical risk	(h_1) Construction technology risk
		(h_2) Equipment installation technology risk
(E) Environment	(i) Positive effect	(i_1) Pollutant emission reduction
		(i_2) Energy saving effect
	(j) Negative effect	(j_1) Sound pollution
		(j_2) Air Pollution
		(j_3) Water pollution

3 EGS power generation investment program optimization technical route

The technical route for the investment program optimization is based on the investment decision-making process of the EGS power generation project. The entire technical route is divided into three parts: The determination of alternatives for evaluation, the establishment of indicator systems, the determination of weights, and the determination of optimal plans for investment plans. The specific technical route is shown in Fig. 1. Through the cooperation of these three parts, the scientific sequencing of alternatives can finally be achieved, assisting the investment decision of the hot rock power generation investment company.

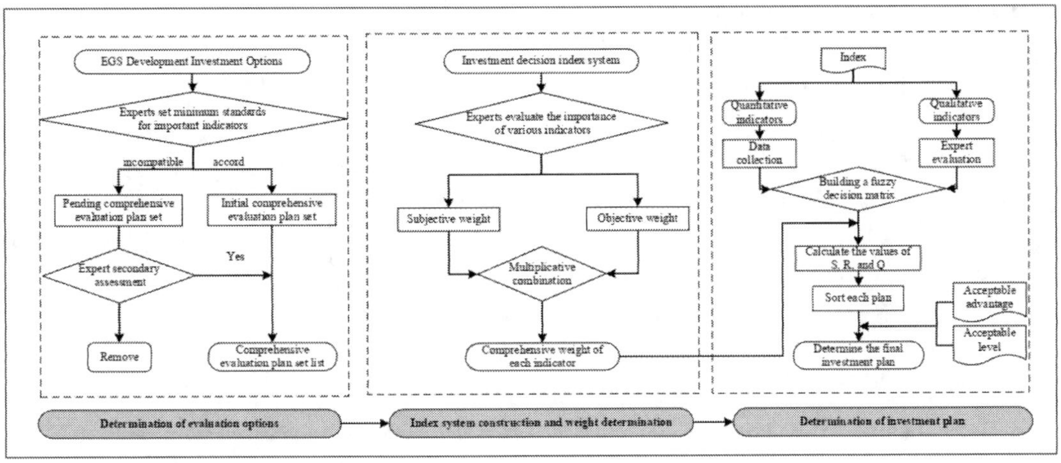

Fig. 1. EGS Power Generation Investment Optimization Technology Roadmap.

3.1 Determination of evaluation options

This article first screens the investment alternatives, and through the relevant screening process, removes some of the investment options from the investment alternatives that do not meet the requirements. The retained investment plans form an evaluation alternative set and enter the subsequent comprehensive evaluation link.

In the first step, the company invited experts from the EGS power generation field to form an expert committee to take charge of this assessment. In the second step, according to the opinions of the enterprise's decision makers and the project feasibility study report, the experts set the minimum standards for some important indicators (in the case of cost indicators, they are the maximum standards). If the value of the relevant indicator is higher than the minimum standard (the cost type indicator is below), the investment plan goes directly to the evaluation candidate set. The third step provides a second opportunity for investment plans that fail to enter the alternative set. Experts analyze the investment value of the project based on the feasibility study report and explore the investment potential of the project. If part of the investment indicators of the project are attractive, they are also allowed to enter the set of evaluation options after consensus of the experts. The fourth step, through these two rounds of assessment, forms a list of options for evaluation. Comprehensive evaluation of the investment plan on the list, select the most appropriate investment plan.

3.2 Determination of index weights based on comprehensive integrated weighting method

Due to the large number of indicators and the weak correlation, this paper intends to adopt a comprehensive integrated weighting method that considers the objective conditions of each item's indicators and also includes the subjective opinions of experts on the importance of each index. The objective weights were determined by the method of entropy weight, and the subjective weights of each index were determined by using the group-based order relation method.

3.2.1 Objective weight calculation

The entropy weight method [12] is a method for determining the weight of an evaluation object based on the amount of information provided by the evaluation index observation value. In the comprehensive evaluation, this paper uses information entropy to obtain the degree of systematic information and evaluates the utility value of information.

According to the entropy method to determine the weight coefficient steps are as follows:

Step 1: Perform dimensionless operations on the attribute values of the j-th index of i-project

$$P_{ij} = \frac{x_{ij}}{\sum_{i=1}^{n} x_{ij}} \quad (1)$$

Step 2: Calculate the entropy of the j-th indicator

$$E(x_j) = -(lnn)^{-1} \sum_{i=1}^{m} P_{ij} \, lnP_{ij} \quad (2)$$

Step 3: Calculate the objective weight of the j-th indicator

When $P_{ij} = 0$, then $E(x_j) = 0$, we can see $0 \le E(x_j) \le 1$, , the degree of deviation of the j index attribute under each scheme $d_j = 1 - E(x_j)$, the objective weight of the j-th index

$$w_j^0 = \frac{d_j}{\sum_{j=1}^{m} d_j} \quad (j = 1,2,\cdots,m) \quad (3)$$

According to the entropy weight calculation formula above, the objective weights of the evaluation indexes of m EGS power generation investment plans are obtained.

3.2.2 Subjective weight calculation

Step 1: determine the order relationship

Each expert individually selects the most important index for the investment decision of EGS power generation projects from the index set $\{X_1, X_2, \cdots, X_m\}$, marked as$X_1^*$, and then picks out the most important indicators from among the unselected indicators. Until all the indicators have been selected, this forms an index of importance: $X_1^* > X_2^* > \cdots > X_m^*$.

Step 2: Determination of the ratio of the importance of each index

The ratio of the importance of the evaluation indicators X_{j-1} and X_j to the expert is denoted as r_k[13]. The table for the assignment ofr_k is shown in Table 2:

$$\frac{w_{j-1}}{w_j} = r_j \ , k = m \ , m-1 \ , m-2 \ , \ \ldots \ , 3 \ , 2 \quad (4)$$

Table 2. r_j assignment table.

r_j	Instructions
1.0	The evaluation index X_{k-1} has the same degree of importance as X_k
1.2	The evaluation index $X_{k-1}$is slightly more important than X_k
1.4	The evaluation index$X_{k-1}$is significantly more important than X_k
1.6	The evaluation index $X_{k-1}$is more important than X_k
1.8	Evaluation index X_{k-1} is more important than X_k

Step 3: Calculation of Group Order Weights

If all experts form a consistent order relationship, the weight calculation formula is

$$\begin{cases} w_m^* = (1 + \sum_{k=2}^{m} \prod_{i=k}^{m} r_i^*)^{-1} \\ w_{j-1}^* = r_j^* w_j^*, \quad j = m,\cdots,2 \\ r_j^* = \frac{1}{L} \sum_{k=1}^{L} r_{kj} \, , j = 2,\cdots,m \end{cases} \quad (5)$$

Where: L represents the number of experts; r_{kj} represents the ratio of expert k to the importance of the indicator $X_{j-1}^*$and X_j^*; w_j^* represents the weight of indicator j. If the ordering relationship formed by experts is inconsistent, assuming the h-order relationship is formed, the weight under each order relationship is calculated according to formula (5), and the final comprehensive weight is obtained through the weighted average method.

$$w_j^1 = \sum_{s=1}^{h} k_s \, w_{js} \quad (6)$$

In the formula: w_{js} represents the weights under s sorting; k_s represents the number of experts forming the s-th sort.

3.2.3 Comprehensive weight calculation

The objective weights w_j^0 and subjective weights w_j^1 are combined by multiplication, and the adjusted comprehensive weights are as follows:

$$\widetilde{w}_j = \frac{w_j^1 \times w_j^0}{\sum_{j=1}^{m} w_j^1 \times w_j^0} \quad (7)$$

3.3 Optimization based on fuzzy VIKOR investment plan

Suppose an EGS power generation project investment multi-criteria problem with K decision makers $L_k(k = 1,2,\cdots,K)$, n alternatives $P_i(i = 1,2,\cdots,n)$and m decision indicators $C_j(j = 1,2,\cdots,m)$.

3.3.1 Initial fuzzy decision matrix

Let the initial fuzzy decision matrix be $X = (\tilde{X}_{ij})_{n \times m}$, which consists of two parts: the quantitative index decision value and the qualitative index decision value.

Table 3. Linguistic ratings and fuzzy numbers.

Linguistic Term (criteria)	Linguistic Term (alternative)	Fuzzy Triangular Number
Very Low	Very poor (VP)	(1, 1, 3)
Low	Poor (P)	(1, 3, 5)
Medium	Fair (F)	(3, 5, 7)
High	Good (G)	(5, 7, 9)
Very High	Very Good (VG)	(7, 9, 9)

Qualitative indicators are obtained through expert scoring. Experts use linguistic variables to score the qualitative indicators of alternatives. Then based on the relationship in Table 3, the scoring results will be converted into triangular fuzzy numbers. Suppose that the triangular fuzzy number $\tilde{X}_{ij}^k = (x_{ij1}^k, x_{ij2}^k, x_{ij3}^k)$represents the decision value of the kth expert on the jth qualitative indicator of the i th equipment scheme, and λ_k represents the k-th expert decision weight $\tilde{X}_{ij} = (x_{ij1}, x_{ij2}, x_{ij3})$ represents the index decision value of multiple experts.

$$\begin{cases} x_{ij1} = \sum_{k=1}^{k} \lambda_k x_{ij1}^k \\ x_{ij2} = \sum_{k=1}^{k} \lambda_k x_{ij2}^k \\ x_{ij3} = \sum_{k=1}^{k} \lambda_k x_{ij3}^k \end{cases} \quad (8)$$

For quantitative indicators, the real data is converted to triangular fuzzy numbers through fuzzification. Conversion method is

$$\begin{cases} x_{ij1} = d_{ij}(1-\alpha_j) \\ x_{ij2} = d_{ij} \\ x_{ij3} = d_{ij}(1+\beta_j) \end{cases} \quad (9)$$

In the formula: d_{ij} represents the real value of the j-th quantitative index of the i-th alternative; α and β represent the fluctuation coefficient of the data downwards and upwards in an uncertain environment, whose value is determined by the expert.

3.3.2 Initial fuzzy matrix standardization

Since the original quantitative index values have different dimensions, they cannot be directly compared and standardization is required before evaluation. After the normalized matrix is $A = (\tilde{A}_{ij})_{n\times m}$, the standardized method is

$$\tilde{A}_{ij} = \begin{cases} \left(\frac{x_{ij1}}{x_{j3}^+}, \frac{x_{ij2}}{x_{j3}^+}, \frac{x_{ij3}}{x_{j3}^+}\right), & c_i \in \Omega_b \\ \left(\frac{x_{j1}^-}{x_{ij3}}, \frac{x_{j1}^-}{x_{ij2}}, \frac{x_{j1}^-}{x_{ij1}}\right), & c_i \in \Omega_c, x_{j1}^- \neq 0 \\ \left(1-\frac{x_{ij3}}{x_{j3}^+}, 1-\frac{x_{ij2}}{x_{j3}^+}, 1-\frac{x_{ij1}}{x_{j3}^+}\right), & c_i \in \Omega_c, x_{j1}^- = 0 \end{cases} \quad (10)$$

Where: $x_{j3}^+ = max_i\{x_{ij3}\}$; $x_{j1}^- = min_i\{x_{ij1}\}$; $\Omega_b$represents a profitable index; Ω_c represents a cost index.

3.3.3 Ordering and optimization of investment programs

The purpose of this section is to sort the alternatives by a fuzzy multi-criteria compromise solution ranking method and make decisions according to the corresponding regulations.

Step1: determine the positive and negative fuzzy ideal solution of each index

Positive fuzzy ideal solution

$$\tilde{F}_j^+ = (a_{j1}^+, a_{j2}^+, a_{j3}^+) \quad (11)$$

In the formula: $a_{j1}^+ = max_i\{a_{ij1}\}$; $a_{j2}^+ = max_i\{a_{ij2}\}$; $a_{j3}^+ = max_i\{a_{ij3}\}$.

Negative fuzzy ideal solution

$$\tilde{F}_j^- = (a_{j1}^-, a_{j2}^-, a_{j3}^-) \quad (12)$$

In the formula: $a_{j1}^- = min_i\{a_{ij1}\}$; $a_{j2}^- = min_i\{a_{ij2}\}$; $a_{j3}^- = min_i\{a_{ij3}\}$.

Step 2: Calculate the fuzzy distance

The vertex method is used to determine the distance between two triangular fuzzy numbers.

$$D(\tilde{F}_j^+, \tilde{U}_{ij}) = \sqrt{\tfrac{1}{3}\left[(a_{j1}^+ - a_{ij1})^2 + (a_{j2}^+ - a_{ij2})^2 + (a_{j3}^+ - a_{ij3})^2\right]} \quad (13)$$

$$D(\tilde{F}_j^+, \tilde{F}_j^-) = \sqrt{\tfrac{1}{3}\left[(a_{j1}^+ - a_{j1}^-)^2 + (a_{j2}^+ - a_{j2}^-)^2 + (a_{j3}^+ - a_{j3}^-)^2\right]} \quad (14)$$

Step 3: Calculate group effect value S_i and individual regret value R_i

$$S_i = \sum_{j=1}^{m} w_j\, D(\tilde{F}_j^+, \tilde{U}_{ij}) / D(\tilde{F}_j^+, \tilde{F}_j^-) \quad (15)$$

$$R_i = max_i\left[D(\tilde{F}_j^+, \tilde{U}_{ij}) / D(\tilde{F}_j^+, \tilde{F}_j^-)\right] \quad (16)$$

Step 4: Calculate the comprehensive index value of multiple criteria and multiple compromises

$$S^+ = min_i\{S_i\} \quad S^- = max_i\{S_i\} \quad (17)$$

$$R^+ = min_i\{R_i\}\ ,\ R^- = max_i\{R_i\} \quad (18)$$

$$Q_i = v\frac{S_i - S^+}{S^- - S^+} + (1-v)\frac{R_i - R^+}{R^- - R^+} \quad (19)$$

Where v denotes the coefficient of the decision mechanism, v>0.5 denotes the decision based on the group effect, v>0.5 denotes the decision based on the individual regret criterion, v =0.5 denotes that the group effect and individual regrets are taken into account simultaneously to make the decision [9-11].

Step 5: Determine the compromise

According to the ascending ordering of S_i, R_i, and Q_i values for each device scheme, let B^1 and B^2be the top 2 rankings after sorting by Q_i values. If B^1 satisfies the following two conditions, it is a trade-off solution.

Condition 1: Acceptable advantages

$$Q(B^2) - Q(B^1) \geq 1/(n-1) \quad (20)$$

Where n is the number of evaluation alternatives.

Condition 2: Acceptable stability in the decision-making process

B^1 must also be ranked first in the order of S_i, R_i. If there is a condition that cannot be satisfied, a series of tradeoffs will be generated, including the following two conditions:

Case 1: If only condition 2 cannot be met, B^1 and B^2 are tradeoff solutions;

Case 2: If the condition is not satisfied, B^1, B^2, ..., B^N are all trade-off solutions, and B^N is satisfied by the relation$Q(B^N) - Q(B^1) < 1/(n-1)$ The maximum N value is determined.

4 Example analysis

A new energy investment company plans to invest in the EGS power station project in northwestern China. Through the previous research, a number of potential investment plans have been formed. The following multi-criteria decision-making is used to select the EGS power generation investment with high value.

4.1 Determination of evaluation options

According to the company's strategy and regional hot rock development and utilization policy, the company has identified five potential investment plans, labeled T_1, T_2, T_3, T_4 and T_5. The company invited five experts in the field of investment in the hot rock power generation project to optimize this plan. After the minimum standards and expert second evaluation, the final investment plans T_1, T_3 and T_4 carry out the subsequent comprehensive evaluation, renumbered as D_1, D_2 and D_3, respectively.

4.2 Calculation of indicator weights

After the assessment of the options was finalized, the Committee of Experts began to calculate the index weights. First, the expert committee evaluates each of the three alternatives based on the project feasibility report, and calculates the objective weights based on the entropy method. Secondly, five experts ranked the importance of each index in the indicator system, forming a total of 1 order relations, 2 order relations 2, and 2 order relations 3, and calculated subjective weights based on the order relation method. Finally, the subjective weights and objective weights are calculated by the formula of the multiplicative combination and the comprehensive weights are calculated. The specific weights are shown in Table 4.

4.3 Optimization of investment program

4.3.1 Obtaining and standardizing fuzzy decision data

First, each secondary index is classified by nature and type. Indicators a_1, a_2, b_1, c_1, e_1-e_3, and f_1-f_3 are quantitative indicators, and others are qualitative indicators. In the same way, indicators e_1-e_3, f_1, f_2, g_1, g_2, h_1, h_2, j_1-j_3 are cost indicators, and others are profit-oriented indicators.

Second, due to the different nature of the secondary indicators, different methods are needed to obtain their fuzzy decision values. Experts (S_1, S_2, S_3, S_4, S_5) use linguistic variables to score the qualitative indicators of each investment plan. Then the expert scoring results are converted into triangular fuzzy numbers according to Table 3, and the fuzzy decision value of each expert's qualitative indicator is obtained, as shown in Table 5. According to formula (8), each expert's decision value is summarized and the comprehensive decision value of the qualitative indicator is obtained.

After the determination of the qualitative indicator decision value is completed, the expert committee needs to negotiate the fluctuation coefficient of each quantitative indicator. The upward volatility coefficient indicates the magnitude of the growth of the decision value under uncertainty, and the downward fluctuation coefficient indicates the magnitude of the decline of the decision value. Obtain specific quantitative decision-making values from the project feasibility study report, and then convert the real data into triangular fuzzy numbers according to formula (9) to obtain quantitative fuzzy decision values. Finally, according to formula (10), the fuzzy decision data is standardized so that the decision values of each index are comparable. The standardized data is shown in Table 6.

Table 4. Determination of subjective and objective weights.

Guidelines	First-level indicators	Secondary indicators	w_j^0	w_j^1	$\tilde{w}_j$
Advantage（0.3212）	Resources（0.2312）	a_1	0.0802	0.0921	0.1443
		a_2	0.0625	0.0712	0.0869
	Location（0.0900）	b_1	0.0186	0.0164	0.0060
		b_2	0.0132	0.0104	0.0027
		b_3	0.0681	0.0612	0.0814
Quality（0.2898）	EGS system（0.2716）	c_1	0.0754	0.0715	0.1053
		c_2	0.0528	0.0422	0.0435
		c_3	0.0423	0.0315	0.0260
		c_4	0.0603	0.0822	0.0968
	Construction（0.0182）	d_1	0.0022	0.0087	0.0004
		d_2	0.0286	0.0319	0.0178
Economic（0.2301）	Investment（0.1111）	e_1	0.0674	0.0717	0.0944
		e_2	0.0321	0.0218	0.0137
		e_3	0.0137	0.0115	0.0031
	Benefit（0.1189）	f_1	0.0627	0.0705	0.0863
		f_2	0.0372	0.0381	0.0277
		f_3	0.0188	0.0134	0.0049
Risk（0.0649）	Social risk（0.0487）	g_1	0.0501	0.0429	0.0420
		g_2	0.0211	0.0163	0.0067
	Technical risk（0.0162）	h_1	0.0259	0.0118	0.0060
		h_2	0.0177	0.0295	0.0102
Environment（0.0940）	Positive effect（0.0298）	i_1	0.0301	0.0289	0.0170
		i_2	0.0225	0.0292	0.0128
	Negative effect	j_1	0.0288	0.0231	0.0130

	(0.0642)	j_2	0.0259	0.0243	0.0123
		j_3	0.0418	0.0477	0.0389

Table 5. Fuzzy decision values of qualitative indicators.

Index	D_1					D_2					D_3				
	S_1	S_2	S_3	S_4	S_5	S_1	S_2	S_3	S_4	S_5	S_1	S_2	S_3	S_4	S_5
b_1	M	M	H	M	H	M	L	M	H	M	H	L	H	M	M
b_3	L	M	M	M	H	H	M	M	M	M	M	M	L	M	H
c_2	H	M	L	M	L	L	H	L	L	M	L	M	L	M	L
c_3	L	VL	VL	L	M	L	L	M	M	L	M	L	M	L	L
c_4	H	VH	H	VH	M	H	M	H	VH	H	H	M	H	H	VH
d_1	VH	M	H	H	VH	M	H	VH	H	M	VH	H	H	H	M
d_2	H	L	H	M	H	H	H	M	H	H	H	H	M	M	H
g_1	H	H	M	H	L	L	H	H	H	M	H	H	L	H	VH
g_2	H	M	L	H	M	H	H	L	M	VH	VH	L	M	H	H
h_1	M	M	H	M	H	M	H	M	H	L	H	H	M	M	M
h_2	VH	VH	H	VH	M	VH	VH	H	H	H	M	M	H	H	VH
i_1	H	M	L	H	M	H	H	H	M	L	H	M	VH	L	M
i_2	VH	VH	H	M	H	H	M	VH	H	H	VH	H	L	H	H
j_1	H	M	H	H	VH	VH	H	H	H	M	H	VH	H	VH	H
j_2	H	H	M	VH	H	H	VH	M	H	H	H	H	M	H	H
j_3	M	L	M	VL	VL	L	M	VL	VL	M	M	L	L	VL	L

Table 6. Decision values after standardization for each scenario and Fuzzy distance calculation table.

Index	D_1	D_2	D_3	$D(\widetilde{F}_j^+, \widetilde{A}_{1j})$	$D(\widetilde{F}_j^+, \widetilde{A}_{2j})$	$D(\widetilde{F}_j^+, \widetilde{A}_{3j})$	$D(\widetilde{F}_j^+, \widetilde{F}_j^-)$
a_1	(0.62,0.78,0.83)	(0.81,0.94,1.00)	(0.72,0.86,0.91)	0.17	0.00	0.09	0.17
a_2	(0.85,0.94,1.00)	(0.71,0.85,0.92)	(0.79,0.88,0.95)	0.00	0.11	0.06	0.11
b_1	(0.75,0.80,0.86)	(0.82.0.88,0.95)	(0.87,0.92,1.00)	0.13	0.05	0.00	0.13
b_2	(0.83,0.91,1.00)	(0.65,0.79,0.91)	(0.70,0.86,0.94)	0.00	0.14	0.09	0.14
b_3	(0.68,0.82,0.92)	(0.72,0.89,1.00)	(0.68,0.82,0.93)	0.07	0.00	0.06	0.07
c_1	(0.45,0.66,0.83)	(0.56,0.78,0.94)	(0.62,0.83,1.00)	0.17	0.06	0.00	0.17
c_2	(0.81,0.90,1.00)	(0.62,0.75,0.86)	(0.68,0.82,0.89)	0.00	0.16	0.11	0.16
c_3	(0.55,0.67,0.83)	(0.68,0.75,0.87)	(0.79,0.86,1.00)	0.20	0.12	0.00	0.20
c_4	(0.72,0.85,0.95)	(0.66,0.79,0.91)	(0.81,0.89,1.00)	0.06	0.12	0.00	0.12
d_1	(0.78,0.87,0.96)	(0.81,0.90,1.00)	(0.72.0.82,0.93)	0.03	0.00	0.08	0.08
d_2	(0.39,0.49,0.72)	(0.51,0.60,0.78)	(0.59,0.71,1.00)	0.24	0.15	0.00	0.24
e_1	(0.23,0.41,1.00)	(0.15,0.31,0.55)	(0.21,0.35,0.68)	0.00	0.27	0.19	0.27
e_2	(0.81,0.86,0.91)	(0.83,0.92,1.00)	(0.80,0.82,0.90)	0.15	0.00	0.08	0.15
e_3	(0.77,0.83,0.96)	(0.81,0.88,1.00)	(0.74,0.81,0.89)	0.04	0.00	0.09	0.09
f_1	(0.71,0.83,0.93)	(0.84,0.95,1.00)	(0.81,0.89,0.96)	0.11	0.00	0.05	0.11
f_2	(0.76,0.85,0.97)	(0.74,0.82,0.91)	(0.83,0.91,1.00)	0.06	0.09	0.00	0.09
f_3	(0.63,0.81,1.00)	(0.56,0.79,0.91)	(0.55,0.78,0.83)	0.00	0.07	0.11	0.11
g_1	(0.66,0.81,1.00)	(0.61,0.73,0.90)	(0.65,0.78,0.94)	0.00	0.08	0.04	0.08

g_2	(0.28,0.53,1.00)	(0.17,0.44,0.63)	(0.22,0.48,0.71)	0.00	0.23	0.17	0.23
h_1	(0.43,0.62,0.85)	(0.50,0.69,0.87)	(0.56,0.77,1.00)	0.14	0.09	0.00	0.14
h_2	(0.05,0.10,0.38)	(0.08,0.21,1.00)	(0.06,0.16,0.59)	0.36	0.00	0.24	0.36
i_1	(0.54,0.72,0.83)	(0.62,0.77,1.00)	(0.49,0.69,0.81)	0.11	0.00	0.14	0.14
i_2	(0.60,0.82,0.97)	(0.48,0.69,0.89)	(0.60,0.82,1.00)	0.02	0.12	0.00	0.12
j_1	(0.12,0.23,1.00)	(0.09,0.13,0.29)	(0.10,0.18,0.63)	0.00	0.41	0.22	0.41
j_2	(0.15,0.28,0.45)	(0.31,0.43,1.00)	(0.26,0.39,0.63)	0.34	0.00	0.22	0.34
j_3	(0.53,0.67,1.00)	(0.41,0.49,0.61)	(0.50,0.62,0.78)	0.00	0.26	0.13	0.26

4.3.2 Optimization and ranking of investment programs

According to equations (11) and (12), the fuzzy positive ideal solution $\tilde{F}_j^+$ and the fuzzy negative ideal solution $\tilde{F}_j^-$ for each index are determined. Then calculate the fuzzy distance according to equations (13) and (14). The calculation results are shown in Table 6.

According to equations (10), (11) and (14), calculate the maximum group effect S_i, the minimum regret value R_i, and the multi-criteria compromise solution ranking method synthesis index valueQ_i (where v takes 0.5). The calculation results are shown in Table 7.

Table 7. Table of S_i, R_i and $Q_i$values.

Index	D_1	D_2	D_3
S_i	0.60	0.54	0.43
R_i	0.14	0.10	0.08
Q_i	0	0.53	1

Roots S_i, R_i, and Q_i sort the alternatives in ascending order, and select the plan according to the decision criteria. D_1 is the scheme ranked first in ascending order of Q_i value, satisfying $Q(D_1) - Q(D_2) > 0.5$, but does not satisfy condition 2, so D_1 and D_2 are trade-off solutions. Therefore, D_1 and D_2 are the final solutions determined according to the fuzzy multi-criteria compromise solution sorting method. New energy companies can invest in these two options.

5 Conclusion

EGS will be the key clean energy for development and utilization worldwide in the future. This paper constructs an evaluation index system for the EGS power generation investment scheme, which uses the entropy weight method and the order relationship analysis method based on group relations to calculate the comprehensive weight of the index as well. Combining the fuzzy set theory with the multi-criteria compromised sorting method, the optimal technical route for the EGS power generation investment scheme was proposed. Through this technical route, an optimal solution that takes into account the largest group effect and the smallest individual regret can be obtained, and the conflict between the ambiguity of the decision attribute and the index is properly solved.

Using this technical route, the case analysis of an EGS power plant project of a company was conducted. The results of the decision show good accuracy, which proves the validity and practicability of the framework.

Acknowledgment

The authors wish to thank anonymous reviewers for their constructive reviews and suggestions, which have greatly helped us improve this paper. This project was supported by the Fundamental Research funds for Central Universities, China University of Geosciences (Wuhan) (1810491T09) and Laboratory Research Funds, China University of Geosciences (Wuhan) (SKJ2018240).

References

1. Olasolo P, Juárez MC, Olasolo J, et al. Economic analysis of Enhanced Geothermal Systems (EGS). A review of software packages for estimating and simulating costs. Applied Thermal Engineering, 104, 647-658 (2016)
2. Wang Guiling, Zhang Wei, Liang Jiyun, et al. Evaluation of geothermal resource potential in China. Earth Journal, **38**, 448-459 (2017)
3. Zhang Aiguo. The commercialization of power generation from dry hot rock still needs time. China Petrochemical News, 007 (2017)
4. Zeng Yijin. Progress and Thinking on Thermal Energy Development of Dry Hot Rock. Petroleum Drilling Technology, 2, 1-7 (2015)
5. Liu Guiying, Li Jijiang, Zhang Yang. Current Situation and Countermeasures of Geothermal Resources Exploration and Development. China Land and Resources Economics, **23**, 17-20 (2010)
6. Zhuang Qingxiang. Discussion on the development potential, benefit analysis and development planning of geothermal energy resources in dry hot rock. Energy and Environment, 2, 2-3 (2017)
7. Tan Xu, Wu Junjiang, Mao Taitian, Tan Yuejin.Multi-attribute intelligent decision-making based on triangular fuzzy number hesitant intuitionistic fuzzy sets. Systems Engineering and Electronics, **39**, 829-836 (2017)
8. Jiang Wenqi. FVIKOR-based multi-criteria decision-making method for triangular fuzzy

numbers. Control and Decision, **31**, 1330-1334 (2016)

9. Tavana M, Caprio D D, Santos-Arteaga F J. An extended stochastic VIKOR model with decision maker's attitude to risk. Information Sciences, 432, 301-318 (2018)
10. Awasthi A, Govindan K, Gold S. Multi-tier sustainable global supplier selection using a fuzzy AHP-VIKOR based approach. International Journal of Production Economics, 195, 106-117 (2017)
11. Gupta P, Mehlawat MK, Grover N. Intuitionistic fuzzy multi-attribute group decision-making with an application to plant location selection based on a new extended VIKOR method. Information Sciences, s 370–371, 184 -203 (2016)
12. Hu Fang, Liu Zhihua, Li Shubiao. Research on risk assessment of public engineering projects based on entropy method and VIKOR method. Journal of Hunan University(Natural Science), **39**, 83-86 (2012)
13. Wang Shuji. Study on the weight determination and evaluation method of post-composition evaluation index for large-scale irrigation area water saving reform project. Xi'an University of Technology, (2009)

Research on the anti-Plugging Property of Permeable Concrete Pavement

Dong Liu , Xia Liu, *Weiwei* Han, *Jing* Chen

China West Construction Group Southwest Co., Ltd., Chengdu, China

Abstract. Permeable concrete pavement is a kind of porous road, which can allow rainwater to penetrate into the ground and maintain the recyclability of water resources. However, with constantly using, its voids may be blocked with the impurities in the rainwater, such as leaves, silt, etc. If that happens, the permeable function of the permeable concrete pavement will be affected. In this paper, the different structure of the permeable concrete pavement is studied, including the variation trend of the coefficient of permeability in the simulated plugging and the recovery rate of the permeable concrete pavement after cleaning and dredging. The results show that the upper small size coarse aggregate (4.75 mm to 9.5 mm) structure of the porous concrete is conducive to filter out most of the impurities, and it will reinforce the resistance to blocking of permeable concrete. But, it is not easy to recover after blockage, if the upper small size coarse aggregate is too thick. The anti-blocking performance and post-blocking recovery rate of permeable concrete are better, when the upper layer thickness is 15 mm.

1 Introduction

With the promotion of urban construction, the proportion of impervious pavement has been rising constantly. Ordinary concrete pavement consists mainly of proportional cementitious materials (organic, inorganic and organic-inorganic combination), granular aggregate, water, chemical admixtures and mineral admixtures. Ordinary concrete pavement brings convenience to people, which has greatly improved people's living standard and production efficiency, however, [1] this kind of pavement is impervious, and surface flow is hard to penetrate, causing flooding in urban areas. Moreover, the air permeability of ordinary concrete pavement is poor, which is not convenient for heat exchange. When the sun irradiates the ground, the temperature of the earth will rise, and it will lead to urban heat island effect and urban environment deterioration.

Compared with ordinary concrete, the preparation of permeable concrete reduces or [2] avoids the use of fine aggregate, forming intercommunicating pore microstructure. The porous structure of permeable concrete makes the pavement have good permeable performance, and the rain can seep into the ground quickly, which can effectively reduce or eliminate flood disasters caused by urban rainstorms. However, the surface runoff produced by rainfall contains a large number of suspended particles that can flow into the porous structure of permeable concrete pavement along with the water continually, leading to the decrease of permeability, shortened service life of permeable concrete pavement and increasing the occurrence of flood and freeze-thaw disasters in cities. Several measures had been proposed such as infiltration, retention, storage, purification, usage and discharge, and these would guide the promotion of sponge city construction in future.

Compared with conventional concrete, the strength and structure performance of permeable concrete are more variable, which mainly depends on the porosity. According to some literatures, [3] the porosity factors of permeable concrete pavement usually vary among 15% to 25%, and the permeability coefficients mostly range from 2mm/s to 6mm/s, but some of them even reach 10mm/s. The pores in permeable concrete pavement can be blocked by suspended particles, such as sediment and debris in surface runoff produced by rainfall, leading to the reduction of permeability and shortened service life of permeable concrete pavement. Studies [4] have shown that the permeability coefficient of less than 15% of permeable concrete pavement is more than 50mm/h after running for 5-6 years without maintenance. This paper studies the attenuation laws of permeability coefficient for permeable concrete pavement under quickly plugging and slowly plugging, as well as recovery rate through high pressure gun. The results will provide theoretical basis for the design of permeable concrete pavement with anti-plugging property and promote the use of pervious concrete.

2 Materials and procedures

2.1 Materials

Cement: P.O 42.5 produced by Lafarge Cement. The physical performance of the cement used in this article is listed in Table 1.

Aggregate: Artificial broken pebbles with single grading size. The experiment has used two different sizes of coarse aggregate, and the physical performance of coarse aggregate is shown in Table 2.

Water: tap water in Chengdu.

Sand used for clogging: manufactured sand located at Chengdu, gradation: 0~0.15mm、0.15mm~0.3mm、0.3mm~0.6mm、0.6mm~1.18mm and 1.18mm~2.36mm.

2.2 Mixture proportion and specimen preparation

As is shown in Fig.1, two kinds of permeable concrete structure are studied in this paper, and the proportions of samples are listed in Table 3.

Table 1. Physical performance of cement.

Density (g/cm³)	Specific area (m²/kg)	Setting time (min)		Compressive strength (MPa)		Flexural strength (MPa)	
		Initial setting	*Final setting*	*3d*	*28d*	*3d*	*28d*
3.11	374	145	540	27.4	48.9	5.6	8.1

Table 2. Physical performance of coarse aggregate.

Particle size (mm)	Apparent density (kg/m³)	Packing density (kg/m³)
4.75-9.5	2710	1700
9.5-16	2690	1700

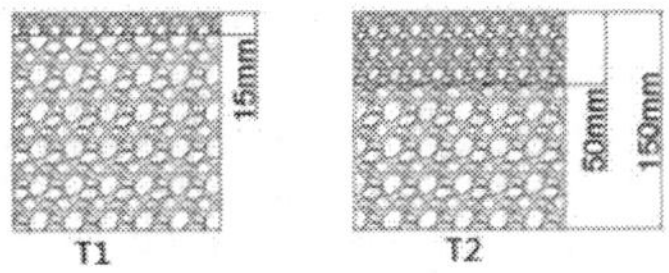

Fig. 1. Structures of permeable concrete pavement.

Table 3. Proportions of permeable concrete samples.

sample	cement (kg)	Coarse aggregate (kg)	Water to binder ratio	Size of aggregate (mm)	Upper crushing pebble thickness(mm)
T1	400	1700	0.28	4.74-9.5 /9.5-16.0	15
T2	400	1700	0.28	4.74-9.5 /9.5-16.0	50

According to literatures, [5] 7 mix proportions of plugging materials are designed in the experiment, as is shown in table 4. A special device is used to test the permeability coefficient, as is shown in Figure.3. The water head height in this device is set to 150mm. After the flow is stable, adding respectively the 7 mix proportions of plugging materials designed to 7 corresponding permeable concrete pavements samples made of the same batch, and reading the flow value every ten seconds to calculate the permeability coefficient in the following formula (1) [6-7], then potting the evolution of the permeable coefficient over time. The selection of the ideal plugging materials must consider that the results of the experiment should be evident and reasonable, so plugging materials sample 1 may be the optimal according to Fig.2. Therefore, sample 1 was used for the following quickly plugging and slowly plugging experiments.

Table 4. Plugging materials designed for permeable concrete pavement.

Sample	1.18-2.36mm(g)	0.6-1.18mm(g)	0.3-0.6mm(g)	0.15-0.3mm(g)	Under 0.15mm(g)
1	0	5	10	10	5
2	0	0	12.5	12.5	5
3	0	0	0	20	10
4	15	15	0	0	0
5	6	12	12	0	0
6	5	8	8	9	0
7	5	5	7	8	5

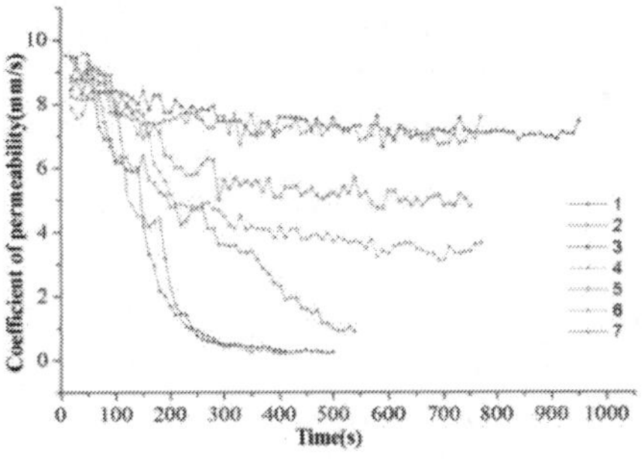

Fig. 2. Attenuation figure of permeability coefficient for permeable concrete pavement under different plugging materials.

2.3 Experiment

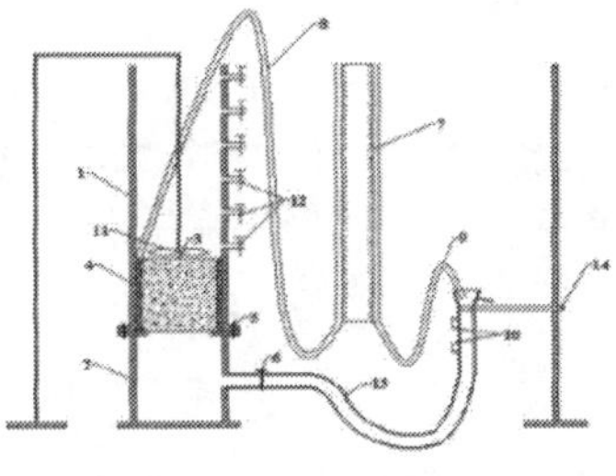

Fig. 3. Permeable coefficient measurement device.
1.Upper canister; 2.lower canister; 3.Permeable concrete pavement sample; 4.Silicone pad; 5.Bolt; 6.Valve; 7.Graduated scale; 8.Rubber hose; 9.Rubber hose; 10.Electromagnetic flow meter; 11.Propeller; 12.Valve; 13.Flexible hose; 14.Clamp.

As shown in Figure 3, a special device was made to order to test the permeability coefficient, which has been used in 2.2 above. This device includes a container, in that a permeable concrete block should be placed. The sand adder is scattered through the propeller (3). The water will enter the container through rubber hose (8), and then penetrates the preamble concrete into the cavity (2). Finally, it flows out of the container through flexible hose (13). The amount of water (Q) passed through the container in a certain time (t) can be known by the graduated scale (7). So the permeability coefficient (K_T) under a certain water pressure can be calculated by formula (1). And the water pressure can be adjusted by throttle valve (12).

$$k_T=\frac{QL}{AHt} \quad (1)$$

K_T-Permeability coefficient at T℃ (mm/s);

Q-The amount of water seeping out during time T (mm^3);

L-Thickness of the specimens (mm);

A-Above-mentioned area of the specimens (mm^2);

H- Water head (mm);

t- Time (s).

The porosity is a common index to evaluate the permeability of permeable concrete. And the initial porosity of permeable concrete pavement samples can be determined by the following equation [8-9]:

$$P=1-(\frac{W_1-W_2}{V_1\rho_w})\times 100\% \quad (2)$$

W_1- Mass of samples after drying in 60℃ oven for 24 hours, kg;

W_2-Mass of samples after immersing in water for 24 hours, kg;

V_1- Value of samples, mm^3;

ρ_w-Density of water, kg/mm^3.

3 Results and analysis

3.1 Measurement of porosity of permeable concrete

The results of porosity test are shown in table 5. The porosity value of permeable concrete T1 is higher than that of permeable concrete T2, because of thicker bottom layer in permeable concrete T1.

Table 5. Measured porosity values of permeable concrete

Sample	Drying mass(kg)	Mass in water(kg)	Porosity (%)	Average value of porosity (%)
T1	6.868	4.215	21.40	20.675
	7.091	4.340	18.50	
	7.035	4.395	21.80	
	7.019	4.350	21.00	
T2	7.023	4.210	16.70	18.375
	7.047	4.295	18.50	
	7.006	4.270	18.90	
	7.120	4.400	19.40	

3.2 Slow plugging test of permeable concrete pavement

Plugging material Z1 was used in plugging test, adding 5g for the interval of 1h, and the water head was set to be 100mm, without stir, to simulate the process of plugging with low speed just as normal rainfall condition. It can be seen from Fig.4 that ladders appear in the evolution curve of the permeable coefficient during the interval of the addition of the clogging material. This phenomenon [10] can be attributed to that unsteady plugging material stuck in permeable concrete can be carried away by current, leading to a transient recovery of permeable coefficient. However, there is a limit to this recovery (Table 6).

3.3 Quickly plugging test of permeable concrete pavement

Plugging material Z2 was used in plugging test, adding 30g for the interval of 90s, and the water head is set to be 100mm, stir at the speed of 30r/min. a numerical value would be readied from the electromagnetic flow-meter to calculate the permeability(Table 7).

Table 6. Mix design of plugging material used for slowly plugging test of permeable concrete

Sample	1.18-2.36mm (g)	0.6-1.18mm (g)	0.3-0.6mm (g)	0.15-0.3mm (g)	Under 0.15mm (g)	Number of times adding sand
Z1	0	0.83	1.67	1.67	0.83	24

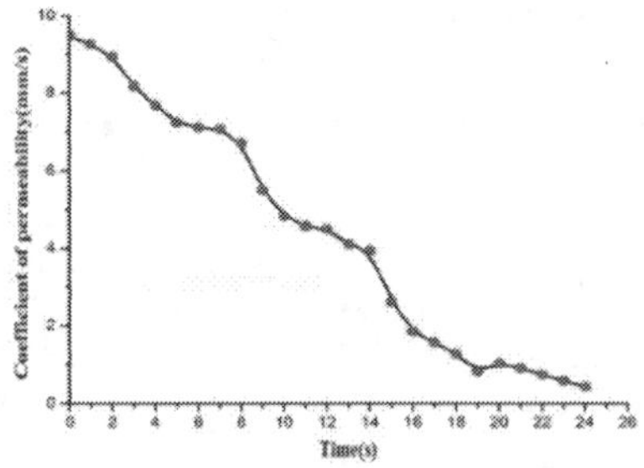

Fig. 4. Attenuation diagram of permeable concrete under slowly plugging.

It can be detected in Fig.5 that the initial permeability coefficient for each sample is different even though the samples were made the same batch. This is because permeable concrete has a porous structure with a large internal surface area, and the manual operation error occurred when the samples formed, which the consistency of the initial coefficient of permeability could not be guaranteed even though their attenuation trends were roughly the same. When the thickness of the surface layer made of 4.74-9.5mm is 15mm, the initial coefficient of permeability is bigger than that of 50mm, which is consistent with the results of the initial porosity of samples.

Compared with slowly plugging test of permeable concrete, there is no obvious that permeable coefficient recovery step on the quickly plugging attenuation diagram. This can be attributed to the short time interval of sand adding for quickly plugging test. The unstable particles could not be washed away by water after each sand addition and stayed in the permeable concrete samples together with the next time adding sand, leading to the disappearance of steps in permeability coefficient attenuation diagram and the sustained decrease of permeability coefficient for permeable concrete [11].

 https://doi.org/10.1051/e3sconf/20185701002

Table 7. Mix design of plugging material used for quick plugging test of permeable concrete

Sample	1.18-2.36mm (g)	0.6-1.18mm (g)	0.3-0.6mm (g)	0.15-0.3mm (g)	Under 0.15mm (g)	Number of times adding sand
Z2	0	5	10	10	5	24

3.4 Recovery test of permeability coefficient of permeable concrete

In recent years, many researchers [12] have studied the recovery of permeability coefficient of permeable concrete pavement. The main measures for improving the permeability of permeable concrete are broom sweeping, high pressure water guan washing and vacuum suction etc. After the comparing of several maintenance measures for permeable concrete pavement, researchers [13] point out that the high pressure gun is more effective than other measures. In this paper, the permeability coefficients of permeable concrete samples are restored by high pressure gun to study the recovery rate of the two permeable concrete pavements.

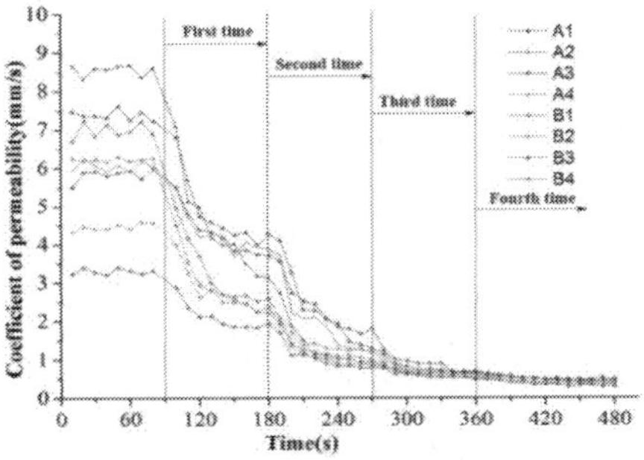

Fig. 5. Attenuation diagram of permeable concrete under quickly plugging.

The recover values and recovery rates of the samples are shown in Table 8 after the first restore with high pressure gun. From the table data, the recovery capacity of permeable concrete samples whose upper layer is 15mm is better than that of 50mm by 10%. This can be explained by that the thinner the size of the upper layer made of 4.74-9.5mm artificially broken pebbles, the easier to be washed out by high pressure gun.

Table 8. Recovery rate of permeable concrete under high pressure gun

Sample		P_I(mm/s)	P_F(mm/s)	$\rho_{(\%)}$
A	A1	8.544	7.141	83.58
	A2	6.044	4.968	82.20
	A3	7.378	6.003	81.36
	A4	8.094	6.748	83.37
B	B1	6.969	4.716	67.67
	B2	4.453	3.541	79.52
	B3	5.822	4.506	77.40
	B4	7.797	5.768	73.97

P_I-Initial coefficient of permeability;
P_F-Coefficient of permeability after primarily recovered test;
ρ - Degree of recovery, $\rho = (P_I/P_T)\times 100\%$

Table 9 shows the recovery and the permeable coefficients of samples after repeated plugging test. Based on the data in Table 8, the recovery rate of permeable concrete samples with 15mm upper layer are better than those blocks whose upper layer thickness are 50mm. This theoretically shows that reasonable thickness of the upper layer of the permeable concrete, to some extent, will be benefit to the recovery of the permeability coefficient of permeable concrete pavement, and prolong the life of the permeable concrete pavement.

Table 9. Recovery of permeable concrete sample after repeated plugging test

h(mm)	P_I(mm/s)	P_F(mm/s)	P_S(mm/s)	P_T(mm/s)	$\rho_{(\%)}$
15	7.38	6.00	5.05	5.66	76.7
50	6.22	3.84	3.39	2.75	44.2

h- Thickness of upper layer;
P_I-Initial coefficient of permeability;
P_F-Coefficient of permeability after primarily recovered test;
P_S- Coefficient of permeability after second recovered test;
P_T- Coefficient of permeability after third recovered test;
ρ - Degree of recovery, $\rho = (P_I/P_T)\times 100\%$

4 Conclusion

a) When the unit weight of the samples are same, the thinner the size of the upper layer, the larger the porosity of permeable concrete samples;

b) Due to artificial molding, initial coefficients of permeability of samples are different. The thinner the size of the upper layer is, the larger the initial permeable coefficient will be, and the higher recovery rate the permeable coefficient will have after plugging tests;

c) Steps on the attenuation diagram of permeable concrete fall slowly under slowly plugging test because of sustained flow, while there have fast attenuation rates on the permeable coefficient attenuation diagram;

d) Reasonable thickness of the upper layer will postpone the plugging and prolong the life of permeable concrete pavement, and the upper structure of 15mm would be a priority selection.

References

1. ZOU R, ZHOU W J, and LI X Y, "The generation and research method of urban heat island effect," Rural Economy and science technology, vol. 18, pp.107-113, March 2007.(In Chinese)
2. SIRIWARDENE N R, DELETIC A, and FLETCHER TD., "Clogging of storm-water gravel infiltration systems and filters: insights from a laboratory study," Water Research, vol 41, pp.1433-1440, July 2007.
3. TAN S, FWA T F, and HAN C T, "Clogging evaluation of permeable bases ," Journal of transportation engineering, vol 129, pp.309-315, March 2003.
4. Drake J, Bradford A, and Seters T V, "Hydrologic performance of three partial-infiltration permeable pavements in a cold climate over low permeability

 https://doi.org/10.1051/e3sconf/20185701002

soil," Journal of Hydrologic Engineering, vol 19, pp.159-169, September 2014.

5. LIU Y Y, JIANG C, LIU J, LIU H, WU Y X, and ZHAO J H, "Advances on the research of plugging and recovery of permeable concrete pavement," Environment Science &Technology, vol 39, pp.159-163, May 2016.
6. Tennis Paul D, Liming Michael L., and Akers David J., Pervious concrete pavements, Skokie, IL: Portland Cement Association, 2004.
7. Hendricks IL, Noiseless concrete pavements, Retrieved from http://www.eupave.edu/documents/graphics/inventory-of-documents/febelcempublicaties/noiselessconcrete-pavements.pdf, 1998.
8. Crouch LK, Cates MA, Dotson VJ, Honeycutt KR, and Badoe D, "Measuring the effective air void content of Portland cement pervious pavements," Cen Concr Aggr, vol 25, pp.16-20, January 2003.
9. MONTES F, VALAVALA S, and HASELBACH L, "A new test method for porosity measurements of Portland cement pervious concrete," Journal of ASTM International, vol 2, pp.1-13, February 2005.
10. ZHENG M L, WANG B G, and HU C S, "Study on fatigue properties of permeable concrete ," Journal of Chinese highway, vol 17, pp.7-11, January 2004.(In Chinese)
11. CUI X Z, ZHANG J, HUANG D, JIN Q, and HOU F, "Simulation of the rapid plugging test of permeable concrete pavement under heavy rain," vol 29, pp.1-19, October 2016.
12. Henderson V, and Tighe S L, "Evaluation of pervious concrete pavement permeability renewal maintenance methods at field sites in Canada ," Canadian Journal of Civil Engineering, vol 38, pp.1404-1413, December 2011.
13. Chopra M, Kakuturu S, and Ballock C, "Effect of rejuvenation methods on the infiltration rates of pervious concrete pavements," Journal of Hydrologic Engineering, vol 112, pp.426-433, June 2010.

Implementation of a cattle manure biodigester for the production of gas for single family use

AS Mullo[1], *WE* Sanchez[1], and *FW* Salazar[1] *JM* Chacha[2] and *A.* Flores[2]

[1] Universidad de la Fuerzas Armadas - ESPE Departamento de Eléctrica y Electrónica Sangolquí-Ecuador
[2] Cotopaxi Technical University Electromechanical Engineering

Abstract. Cattle manure has an energy potential that is unknown to farmers. In the rural areas of the city of Latacunga - Ecuador there are inconveniences for the transport of the G.L. (Liquefied Petroleum Gas). For this reason, this document shows the implementation of a biodigester to obtain biogas for domestic use at Hacienda Terán located in La Dolorosa Belisario Quevedo, the cattle manure is analyzed, obtaining 13.7% of total solids. In addition, it was established that 100 kg of mixture with the quantification of manure in a period of 12 days. In order to maintain the elevated temperature a greenhouse was built obtaining an average of 28.48°C in a time for the generation of biogas in 30 days and a minimum temperature of 22.7°C. Temperatures that are within the limits established for the generation of biogas. With these parameters, a 4.25 m3 bio-bag was dimensioned generating a production of 1.1 m3 of gas that satisfies the family use of 0.7 m3 Biogas and G.L.P. firing tests were carried out up to 80° C obtaining a calorific value of 4800 kcal. per m3.

1 Introduction

The use of fossil fuels is increasing daily oil gas and coal reserves are increasing at the rate of 600 million barrels (Mb). 400 billion cubic feet (BCF) and 19.2 Giga tons of oil equivalent (GTOE) per year [4]. Especially the use of G.L.P. (Liquefied Petroleum Gas) is escalating with a daily energy demand which is the reason why there are environmental threats in recent years harmful to the welfare of humanity and the planet which include climate change global warming[1-2] pollution of the earth air and rivers polluted by animal manure pathogens. In rural areas it is difficult to transport G.L.P. for domestic use.

The use of alternative energies has a great impact for the care of the environment. in Ecuador a change of the energy matrix[3] has been implemented as an alternative to reduce environmental pollution and the rational use of energy. In order to treat solid bovine waste. the use of a Biodigester. The use of alternative energies has a great impact for the care of the environment. in Ecuador a change of the energy matrix[3] has been implemented as an alternative to reduce environmental pollution and the rational use of energy. In order to treat solid bovine waste. the use of a Biodigester has as its main function to minimize the contamination produced by cattle manure, rivers and lands where they are spilled. in addition to rural areas where the Biodigester is implemented has energy autonomy guaranteeing the quality of life. Biogas is not considered a common business source [4] but its impact is multiple as it can be used simultaneously in the production of electrical and thermal energy and as a fertilizer [5]. as shown in figure1.

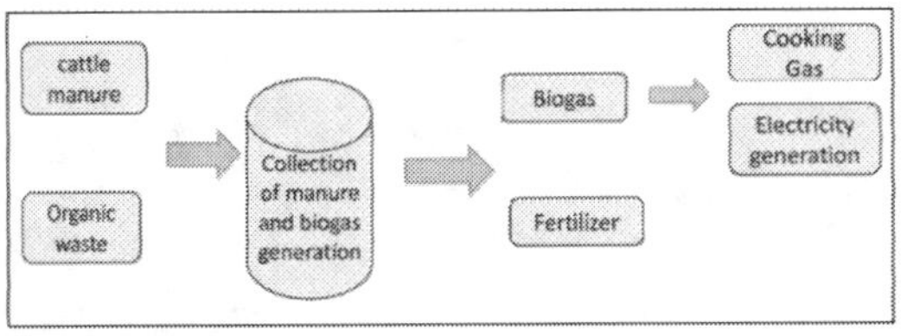

Figure. 1. Use of cattle manure for biogas and fertilizer generation.

In the rural area of La Dolorosa Belisario Quevedo (Latacunga- Ecuador) is a cattle ranch that has had an increase in population with it the need for fuels for domestic use. In this town it is difficult to transport G.L.P so deforestation has increased to obtain firewood and cattle is spilled in the fields and rivers of the place and others burn wood for cooking which has generated pollution problems and increased disease. This is why it is necessary to carry out research to reduce deforestation and diseases as well as to make good use of manure, it takes 0.8 m3 of GLP gas to satisfy the single family demand.

1.1 The Biodigester

A biodigestrer is a closed and hermetic container, inside which the organic substance substance is placed in combination of water – manure, for a time and at an average temperature of 30°C allows anaerobic degradation generating methane gas and fertilizers thus decreasing the polluting potential of excrement [6]. The biodigester can be constructed of brick cement and plastic forming an impermeable container that allows the generation of biogas

1.2 Flexible structure digesters (polyethylene balloon plants)

This type of plant has in the upper part a bag digester in which gas is stored. the entrance and exit are in the same surface of the bag. Its advantages are low cost, easy transportation, little construction technology, high digestion temperatures. easy cleaning. maintenance and emptying. its disadvantages are its short lifespan. high receptivity to damage. low generations[6] [7].

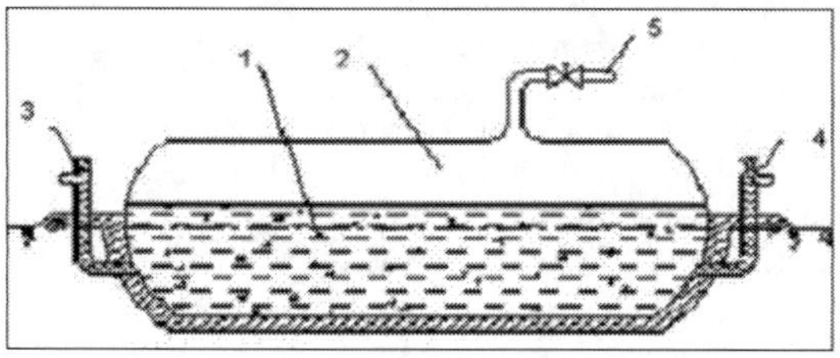

Fig. 2. Use of cattle manure for biogas and fertilizer generation.

Figure 2 shows a biodigester of common use in Andean zones with a protective structure. its components are: (1) Digester. (2) biogas container. (3) manure-water inlet. (4) Bio-fertilizer outlet and (5) gas outlet pipe.

1.3 Chemical composition of cattle manure.

derived from the digestive tract of animals. containing undigested residues of food. and digestive factors such as enzymes. gastric juices. pancreatic and dead cells of the intestinal mucous membrane. living and dead bacteria of the colon and waste products of animal metabolism. Within the chemical composition. manure is not just faecal matter. They are by-products of livestock production that include animal excrement. bedding. washing water. splashed feed. cleaners and hairs. Their composition varies between very large limits. depending on age. class and characteristics of the animals. quantity and digestibility of fodder. concentrated feed consumed by livestock. quantity and type of bed. duration. form of storage and method of handling manure. as shown in table 1[8].

Organic matter	Nitrogen (N)	Phosphorus (P)	Potassium (K)	Magnesio (Mg)	Humidity %
36,1	1,51	1,20	1,51	0,53	25,5

Table 1. Chemical composition of sterol by mass (%).

1.4 Anaerobic digestion process

Anaerobic digestion has four phases to generate biogas, which is the consequence of the degradation of microorganisms, thus forming the gas with a high methane content. In the substract (generic term to designate, in general, the food of microorganisms) there are 4 types of microorganisms that are characterized by their growth speed and according to the continuous degradation for example: the formation of H2, acetic acid or ammonia produced from amino acid acidogenesis [9], as shown in figure 3.

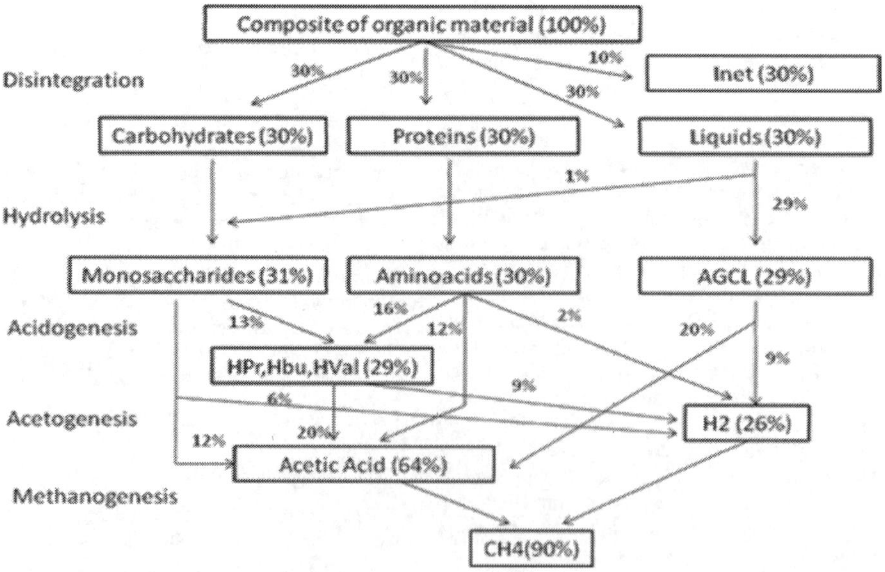

Figure. 2. Stages of anaerobic digestion .

 https://doi.org/10.1051/e3sconf/20185701003

In the methanogenic stage, a large group of strict anaerobic bacteria is generated, which act on the products resulting from the previous stages. Methanogenic microorganisms can be considered as the most important ones within anaerobic microorganisms, since they are responsible for the formation of methane and the elimination of the environment from the products of the previous groups, being, moreover, the ones that give name to the general process of biomethanization. Methanogenic microorganisms complete the anaerobic digestion process by forming methane from monocarbonate substrates or with two carbon atoms bound together by a covalent bond: acetate, H2/CO2,

Table 2. Chemical composition of biogas.

Chemical Compound	Chemical formula	Volumen %
METHANE GAS	CH_4	60-70
CARBON DIOXEDE GAS	CO_2	30-40
HYDROGEN	H_2	1.0
NITROGEN	N_2	0.5
CARBONO MONOXIDE	CO	0.1
OXYGEN	O_2	0.1
SULFHIDRIC ACID	H_2S	0.1

format, methanol and some methylamines [10], as shown in figure 4.

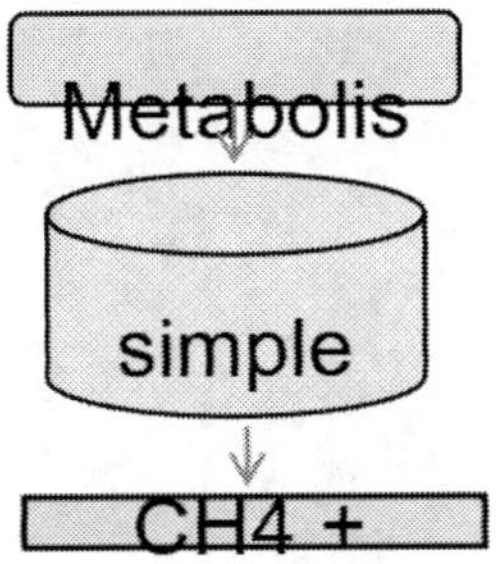

Figure. 4. Stages of metanizacion (methane gas generation) .

1.5 The Biogas.

Biogas is simply the fusion of two main elements, methane and carbon dioxide (CH4 and CO2 respectively), produced by the fermentation of organic matter under anaerobic conditions (absence of oxygen) [5] [11]. It has similar characteristics to natural gas. Around different environments, we find a great variety

Table 3. Chemical composition of biogas.

Day	Daily weight (kg)	Daily weight (lb)	Daily weight (kg)	Daily weight (lb)
1	48	105,8	50,0	110,2
2	49	108,0		
3	51	112,4		
4	50	110,2		
5	52	114,6		
6	49	108,0		
7	52	114,6		
8	48	105,8		
9	52	114,6		
10	49	108,0		

of organic waste such as animal manure, vegetable waste and different domestic waste that processed in a good way would generate biogas. The composition of the biogas is methane and carbon dioxide among other components [12]detailed below table 2:

1.6 The PH index

In order for the biogas generation process to develop satisfactorily. the pH must not fall below 6.0 or rise above 8.0. The pH value in the digester determines not only the biogas production but also its composition. One of the consequences of a fall in pH below 6 is that the biogas generated is very poor in methane and therefore has less energy potential. Since methanogenesis is considered to be the limiting stage of the process. it is necessary to keep the pH of the system close to neutrality. Acidogenic acids are significantly less sensitive to more extreme pH values. as shown in figure 5.

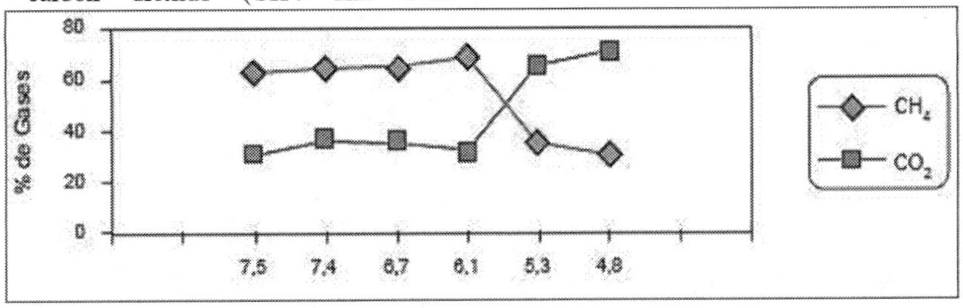

Figure. 5. pH of tuna-guano mixtures

2 Construction of a Polyethinene Bag Bio-Digester

2.1 Quantity of manure production in the stable.

For the design, a field analysis was carried out to determine the daily amount of fresh manure, with a total of 10 Holstein cows weighing approximately 400 kg, resting in a corral at night, the manure was collected with a daily quantity of 20 liters of volume during a period of 10 days at 7:30 a. m.., Table 3 shows that there is a daily production of 50 kg of manure:

2.2 Determination of the daily charge to the Biodigester

Consideration was given to the amount of manure generated daily in the stable, a ratio of 1:1 (1 kg of fresh manure + 1 litre of water), due to the percentage of humidity which is greater than half, as the ratio is equal, then 1kg = 1litre. The density of the manure is almost equal to the density of the water for this reason the mixing ratio is assumed. Then:

VCD=100 kg

VDC: Daily storage volume.

2.3 Dimensioning of the flexible structure polyethylene bag biodigester

To calculate the liquid part corresponding to 75% of the biodigester, the retention time is taken into account, using equation 1 we have:

VL = 100 litres/day*30 days

VL = 3000 litres

VL = 3 m^3

Volume of the gaseous (V.G.) part 25%, and we add a 5% safety factor.

VG = 1200 litres

VG = 1200 litres*m^3/ (1000 litres)

VG = 1.2 m^3

With the results of liquid volume and gaseous volume, it was replaced in equation 3 to obtain the total volume.

VT = 4200 litros

VT = 4200 litros* m^3/(1000 litros)

VT = 4,2 m^3

In order to calculate the length of the Biodigester, the selected site and the availability of a 3 m wide low density tubular polyethylene thermoplastic roll of 3 m wide from equation 4 was taken into account in the stable:

D = 0.95 m

With the diameter, the length required for the Biodigester is calculated from equation 5.

L = 6 m

The dam of the Biodigester was dimensioned based on the length and diameter of the Biodigester, covered all the dam with plastic to avoid landslides, also in the floor it was designed with a difference in level of 2°, with the purpose of the easy flow of the biofertilizer to the irrigation channel.

Li = lower length equal to the length of the Biodigester.

Li = 6 m

The depth is equal to the diameter of the Biodigester:

PD = D.(3/4)

PD = Dyke depth.

PD = 0.71 m

The width of the lower and upper part is equal to the diameter of the Biodigester.

AD = Dyke width.

AD = 0,95

The upper length of the dike is calculated on the basis of the side walls which have an inclination of 45°:

Ls = Upper length.

Ls = 7,42 m

Fig. 6. Construction of the polyethylene digester tank.

2.4 Design and construction of the greenhouse for the biodigester.

The temperature is one of the most important factors in the process of decomposing the mixture inside the biodigester. With the construction of the greenhouse, temperatures ranging from 20° to 35° C were achieved and an average of 28°C, and it also protects the bio-bag from environmental factors such as rain, wind and sun, which allows greater durability.

The dimensions taken into consideration for the construction of the greenhouse are: Length 2.6 meters, width 1.7 meters and height 1.3 meters to keep the area warm, as can be seen in the figure 7.

https://doi.org/10.1051/e3sconf/20185701003

Figure. 7. Greenhouse for the bio-digester of Hacienda Terán.

2.5 Gas Transport Pipe

Once the bio-bag is placed in the slump basin, the transport piping is placed according to the planned layout, with the following materials, the installation can be seen in figures 8 and 9:

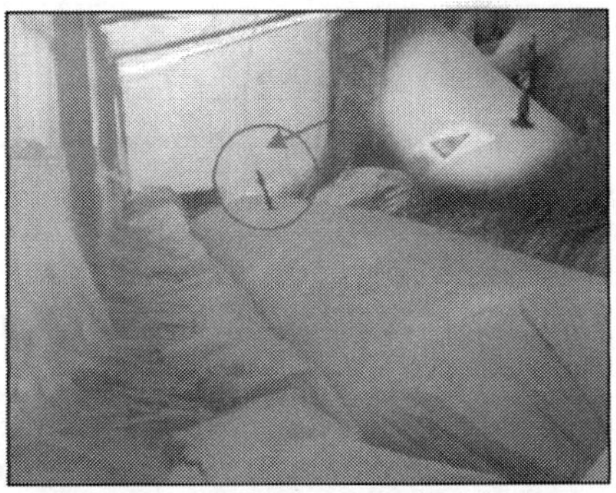

Fig. 8. Biogas output piping from the Biodigester.

Figure. 9. Transport of the biogas by means of a yellow polyethylene pipe.

2.6 Safety valve, moisture filter and sulfuric acid filter

The generation of daily gas creates pressure, it is necessary to avoid that the pressure increases, for this reason at the end of the transport line was implemented a biogas safety valve, a simple exhaust mechanism the same that was made from a 4-litre plastic bottle. The humidity filter was placed to lower the water vapour levels in the biogas, to obtain a cleaner combustion. The biogas has a characteristic smell similar to rotten egg, this is due to a percentage of sulfuric acid contained in the gas, to eliminate sulfuric acid was implemented a steel stamen filter, as can be seen in the figure 10.

Figure. 9. Biogas outlet, filter and valve unit.

3 Analyses of Results Obtained

3.1 Daily and monthly Biogas production.

According to the design characteristics of the Biodigester daily measurements were made to generate the biogas whose daily production is 1050 liters or 1.05 m3. Amount of daily energy produced by the biodigester:

ED = Daily biodigester energy

$$\mathbf{ED} = \frac{1.05 \text{ m3 de biogas}}{\text{day}} \text{ x } \frac{1.2\text{KWh}}{\text{m}^3 \text{ of biogas}}$$

$$\mathbf{ED} = 1.26 \frac{\text{KWh}}{\text{day}}$$

Calculation of monthly energy production:

EM = Monthly energy:

EM = 38.32 KWh / month

Comparative analysis of daily consumption of GLP and biogas.

$$\frac{15 \text{ kg de GLP}}{(1{,}5) \text{ monthy}} * \frac{1 \text{ m}^3 \text{ biogas}}{0{,}45 \text{ kg GLP}} = 22{,}2 \frac{\text{m}^3 \text{ biogas}}{\text{monthly}}$$

$$22{,}2 \frac{\text{m}^3 \text{ de biogas}}{\text{monthly}} * \frac{12 \text{ months}}{1 \text{ year}} * \frac{1\, year}{365\, days} = 0{,}72 \frac{\text{m}^3 \text{ de biogas}}{day}$$

$$0{,}72 \frac{\text{m}^3 \text{ de biogas}}{day} * \frac{1{,}2 \text{ kWh}}{1 \text{ m}^3 \text{ de biogas}} = 0{,}86 \frac{\text{kWh}}{\text{day}}$$

3.2 Characterization of biogas

The characterisation of the energy potential was carried out in gas tests in the Chemistry Laboratory of the National Polytechnic School. And boiling time tests were

 https://doi.org/10.1051/e3sconf/20185701003

performed on one liter of water. with biogas and L.P.G.

Table 4: Chemical characterization of biogas.

COMPONENT	% WEIGHT	% Mol
Nitrogen	4.27	3.97
Oxygen	3.64	2.96
Methane	36.77	59.91
CO_2	54.88	32.51
H2O	0.45	0.65

Table 5: Energetic characteristic of biogas.

Characteristics	OUTCOMES
Relative density	0.904
Average molecular weight (g/mol)	26.07
Calorific value exceeding 25 °C (Btu/lb)	14 294.81
Calorific value exceeding 25 °C (Btu/ft^3)	968.19
Calorific value below 25 °C (Btu/lb)	12 81.56
Calorific value below 25 °C (Btu/ft^3)	872.47

3.3 Temperature and relative humidity in the biodigester

The Figure 11 shows that the construction of the greenhouse maintains the temperature at an average of 28°C. and within the bio-bag it maintains an average of 32°C. this temperature guarantees the biogas generation process. obtaining the following results:

Tmin: 20.1 ° C.
Average = 28.2° C.
Tmax = 34.1 ° C.

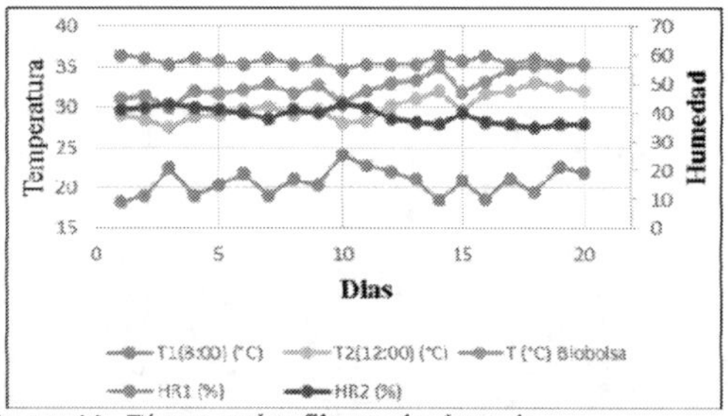

Figure. 11. Biogas outlet. filter and valve unit.

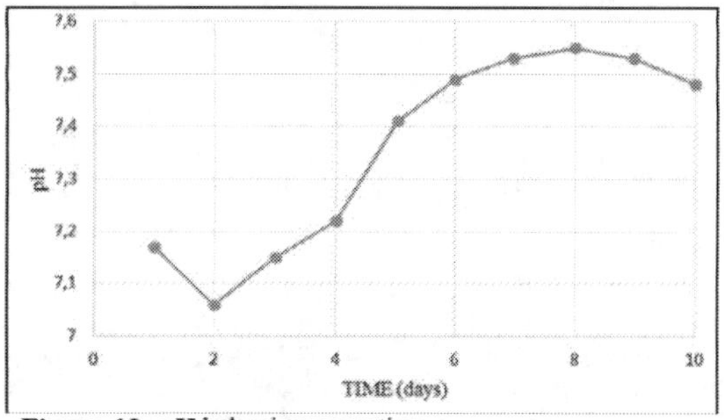

Figure. 12. pH behaviour over time.

Table 6. Temperature and relative humidity inside the greenhouse.

TEMPERATURE AND RELATIVE HUMIDITY INSIDE THE GREENHOUSE					TEMPERATURE INSIDE THE BIOBAG
Day	T1(8:00) (°C)	HR1 (%)	T2(12:00) (°C)	HR2 (%)	T3(12:00) (°C)
1	18,1	60	29	41	31
2	19	59	28,4	42	31,5
3	22,4	57	27,5	43	30
4	19	59	28,9	42	32
5	20,3	58	29	41	31,7
6	21,7	57	29,7	40	32,2
7	19	59	30,1	38	32,9
8	21	57	29	41	31,8
9	20,3	58	29,7	40	32,7
10	24	55	28,1	43	30,7
11	22,7	57	28,3	42	32
12	22	57	30,2	38	33
13	21	57	31	37	33,4
14	18,4	60	32	36	35
15	20,9	58	29,6	40	31,9
16	18,6	60	31,7	37	33,2
17	21,1	57	32	36	34,8
18	19,4	59	33,1	35	35,2
19	22,5	57	32,6	36	35
20	21,8	57	32	36	35,3
	20,66	58	30,1	39	32,77

In order to monitor the correct functioning of the biodigester. daily data was taken from the pH. which is within the established ranges of 6 - 8 for a better biogas quality. which does not need the addition of a substance for pH regulation. the behavior can be observed in the figure 12.

4. Conclusions and Recommendation

4.1 Conclusions

- According to the design of the biodigester. it maintains a gas production of 1.05 m3 of biogas that satisfies the single- family demand of 0.85 m3 of G.L.P. gas. with an average temperature of 28°C and an average pH of 7.3 that generates a methane with good calorific characteristics.
- According to laboratory tests. methane gas has a calorific value of 968.19 (BTU/ft3) and a methane concentration percentage of 59.91 (% moles). which maintain optimal calorific characteristics to be used as a family gas. The time required to heat 2 litres of water with the biogas until reaching 80°C is 8% higher than that of a LPG gas maintains a calorific value of the same characteristics.

4.2 Recommendation

It is necessary to dimension the biodigester according to the family needs that the technical and economic characteristics are analyzed. Disseminate the project to rural areas that have a cattle manure production that allows to generate biogas for single family consumption

References

1. N. Abas. A. Kalair. and N. Khan. "Review of fossil fuels and future energy technologies". Futures. vol. 69. pp. 31-49. May 2015.
2. A. Schmidt. B. E. Law. C. Hanson. and O. Klemm "Distinct global patterns of strong positive and negative shifts of seasons over the last 6 decades" Atmospheric and Climate Sciences. vol. 2 (1). pp. 76 -88. Jan. 2012.
3. M. Castro. "Hacia una Matríz Energética Diversificada en el Ecuador". Centro Ecuatoriano de Derecho Ambiental. vol. 1. pp.15 . (2010)
4. N. Abas. A. Kalair. and N. Khan. "Review of fossil fuels and future energy technologies". Futures. vol. 69. pp. 31-49. May 2015.
5. M. A. Ehsan. C. K. Das. and Mahadi Hasan. "Biogas Based Chain Business: A Road to Sustainable Rural Development". Conference (ICDRET). Jan 2016.
6. R. Botero and T.R. Preston. "Biodigestor de bajo costo para la producción de combustible y fertilizante a partir de excretas" www.produccion-nimal.com.ar/Biodigestores/04-biodigestores.pdf. 1987.
7. T. Arias and L. Lopez Ríos. "Technology proposal for the energy use of the shake bagasse barley from Hatueybrewery".http://scielo.sld.cu/scielo.php?script =sci_artte xt&pid=S2224-61852015000300001. ISSN 2224-618. Santiago de Cuba. 2015.
8. ACPA. "Estiércol bobino Mitos y realidades". http://www.actaf.co.cu/revistas/Revista%20ACPA/2 009/ REVISTA%2004/17%20ESTIERCOL%20BOVINO .pdf Fondo Editorial IDAE. Revista Vol 36. 2009 .
9. Biomasa. "Digestores anaerobios" Fondo Editorial IDAE. Madrid 2007.
10. M. Varnero. "Manual de Biogas". Santiago de Chile. Chile: Viale delle Terme di Caracalla. 2011.
11. W. Pattaraprakorn. R. Saikaew and P. Bhasaputra " Feasible Study of Biogas Energy for Thai Dairy Cooperative from Dairy cow Manure: The Potential and Economic Analysis". Proceedings of the International Conference on Energy and Sustainable Development: Issues and Strategies. ESD 2010.
12. Semarnat and Sagarpa. "Especificaciones tecnicas para el Diseño y Construccion de Biodigestores". Proceedings of the International Conference on Energy and Sustainable Development: Issues and Strategies. Mexico D.F.. Mexico: Firco 2010.

Kalman filter model, as a tool for short-term forecasting of solar potential: case of the Dakar site

A. Mbaye[1],*J.* Ndong[2],*M.L.*NDiaye[1],*M.* Sylla[1],*M.C.* Aidara[1] ,*M.* Diaw[1,3],*M.F.*NDiaye[1],*Papa Alioune* Ndiaye[1] *A.*Ndiaye[1]

[1]Electrical Engineering Department of the Polytechnic High School of Dakar Senegal, Laboratory International Center for Solar Energy, BP 5085 Dakar-Fann, Senegal.
[2]Department of Mathematics and Computer Science, University Cheikh Anta Diop of Dakar Senegal, Laboratory LID.
[3]Department of physical, University Cheikh Anta Diop of Dakar Senegal, Laboratory of Hydraulic and fluid Mechanics.

Abstract. The prediction of solar potential is an important step toward the evaluation of PV plant production for the best energy planning. In this study, the discrete Kalman filter model was implemented for short-term solar resource forecasting one the Dakar site in Senegal. The model input parameters are constituted at a time t of the air temperature, the relative humidity and the global solar radiation. The expected output at time t+T is the global solar radiation. The model performance is evaluated with the square root of the normalized mean squared error (NRMSE), the absolute mean of the normalized error (NMAE), the average bias error (NMBE). The model Validation is carried out by means of the data measured within the Polytechnic Higher School of Dakar for one year. The simulation results following the 20 minute horizon show a good correlation between the prediction and the measurement with an NRMSE of 4.8%, an NMAE of 0.27% and an NMBE of 0.04%. This model could contribute to help photovoltaic based energy providers to better plan the production of solar photovoltaic plants in Sahelian environments.

Keywords. Kalman filter, solar potential, Prediction, Dakar, short-term.

1 Introduction

Senegal has significant potential for the solar resource development (PV and Thermal) with 3000 hours of sunshine per year and an average global irradiation of 5.8 kWh / m 2 / day. The solar photovoltaic market is growing in the world in general and in Africa in particular. Currently, we are witnessing a proliferation of solar photovoltaic plants in Senegal. The integration of these intermittent sources on the SENELEC grid is causing many problems in terms of planning for the maintenance of the balance between consumption and production. Planning is less easy because of the stochastic nature of the renewable energy sources production. A solution can be considered to overcome this limitation : forecasting. Potential prediction can make a substantial contribution in this context by evaluating the penetration rate of intermittent energies on the grid that would limit the risks. The forecasting of the production of intermittent energies makes it possible to anticipate the availability of the sources of production and thus to facilitate the management of the network. The prediction of the solar potential on a horizontal surface can be done in the long term [1], [2] and the short term [3], [4]. It is in this perspective that, this study is made to contribute to the search for reliable prediction methodologies of photovoltaic power. Four methodologies exist in the literature after [5] and [6]. We chose to work with a model based on the kalman filter which has already been widely applied in other sites around the world [7], [8], [9], [10], [11]. The Kalman filter model is convenient for forecasts at different horizons. We want to contribute to the short-term prediction of the solar potential of solar PV plants over the time horizon 20 minutes. Section 2 describes the methodology and tools used, Section 3 presents the results of simulations and their interpretation, and the last section is reserved for conclusion.

2. Materials and Methods

2.1. Study area

Dakar is located in the extreme west of the peninsula of Cape Verde, at the edge of the Atlantic Ocean, between latitude 14 ° 75 N and longitude -17.33 ° W with average altitude of 16m (figure1). It houses some meteorological stations including the measuring station of the International Center for Training and Research in Solar Energy (CIFRES measuring at the Advanced school of Engineering, in Cheikh Anta Diop University, Dakar, Senegal. This station measures global, direct and diffuse solar radiation, ambient temperature, relative humidity, rainfall, wind speed and direction. In this work, we used solar radiation, average temperature as well as relative humidity data from CIFRES station. The data *series* runs from October 1st, 2016 to September 30th, 2017 and measurements are made every 10 seconds. A Campbell Scientific model CR800 datalogger is used.

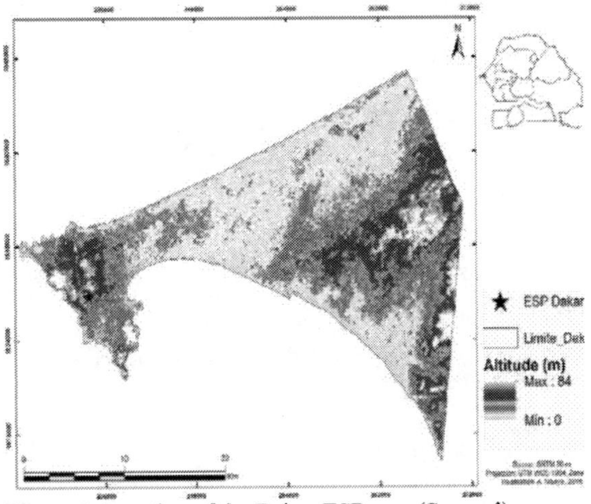

Figure 1. Location of the Dakar-ESP area (Senegal)

2.2. Application of the filter on the study site

We build our prediction model assuming that measured solar power is governed by a hidden signal as in the case of a linear dynamic system following this equation:

$$Y_t = A_t X_t + V_t \qquad (1)$$

$Y_t \in R^m$ denotes a stochastic process containing the measurement variable (observation) at time t, it is composed of m variables corresponding to the measured solar irradiation or photovoltaic power at different points. Generally, due to the intrinsic imperfection of measurement devices, errors occur more often during data collection. V_t denotes a stochastic process representing the measurement error. The stochastic process $X_t \in R^n$ is the hidden state of the system that is transformed into an output by the matrix $A_t \in R^{m \times n}$ that controls the sources of the measurements (solar radiation, ambient temperature, relative humidity). The goal is to find a model for the stochastic process X_t that is able to track solar radiation. We propose a simple stochastic model where X_t at discrete time t, is represented by the linear combination of two components (see equation 2).

$$X_t = C_t \hat{X}_t + \xi_t \qquad (2)$$

$\hat{X}_t$ is a predictable component, ξ_t is a random noise component and $C_t \in R^{n \times n}$,is a parameter matrix describing the dynamics of the state.

Generally, the prediction model can have a structure and the noise process can be generated by any distribution. However, linear stochastic predictive models combined with Gaussian noise have a long record of successful applications in a very wide range of engineering techniques. Our idea is to connect the predictable components with the state model. These variables are not directly observable by measuring devices; we refer to them as hidden states of the system. As we want to estimate the variables X_t as the state variable at the date t, we can use a linear stochastic dynamic system based on state space models.

We then build a temporal model that relates from X_{t+1} to X_t to with the following difference equation (3):

$$X_{t+1} = C_t X_t + B_t U_t + W_t \qquad (3)$$

The $n \times n$ matrix connects the previous state t to the current state $t + 1$;

Compared to our study site, C_t describes the atmospheric conditions since the value of the solar radiation at time t depends on the ambient temperature, the relative humidity. The value of the solar power between t and $t+1$ could change according to these atmospheric conditions. We can therefore estimate C_t given a set of measured data that contains all types of weather conditions to form our forecast model. Indeed, to see the influence of these atmospheric variables on the measurements, it is necessary to take into account the terms B_t and U_t which are deterministic matrices where matrices of parameters describing the dynamic state of $R^{n \times p}$ and R^p *size* ($B_t \in R^{n \times p}$ and $U_t \in R^p$ with p exogenous variables), or W_t is the intrinsic noise process.

To form the complete state space model for the specification of our prediction framework, equations (1) and (3) are now combined to form equation (4). This equation is the classical form of a linear dynamic system with inputs. In this model, we assume that the state noise W_t and the measurement noise V_t are gaussian white noise processes at zero mean that are not correlated with covariance matrices $Q_t \in R^{n \times n}$ and $R \in R^{m \times m}$, respectively. The identification of the parameters C_t, A_t, Q_t and R_t is necessary to properly calibrate the system. The method used for the identification of the parameters is based on the Expectation-Maximization (EM) algorithm described in [7].

The filter also functions as a predictor-corrector algorithm. As an iterative analyzer, he estimates the state of the system using two steps: the forecast intervenes in the phase of updating the time and the correction in the phase of updating the measurement.

2.3. Model Performance Criteria

Within the prediction context of global solar radiation or PV power, there are many methods to check whether a predictor is effective [7]. In this paper, the tools we used are the NRMSE, NMAE, NMBE.

$$NRMSE = \frac{\sqrt{\frac{1}{N}\sum_{i=1}^{N}\left(P_i - \hat{P}_i\right)^2}}{\max(P_i) - \min(P_i)} \qquad (4)$$

$$NMAE = \frac{\frac{1}{N}\sum_{i=1}^{N}\left|P_i - \hat{P}_i\right|}{\max(P_i) - \min(P_i)} \qquad (5)$$

$$NMBE = \frac{\frac{1}{N}\sum_{i=1}^{N}\left(P_i - \hat{P}_i\right)}{\max(P_i) - \min(P_i)} \quad (6)$$

3 Results

3.1 Results obtained

Table 1. Performance of prediction model for time horizon 20 minute.

Model	NRMSE	NMBE	NMAE	Time horizons
EM	0.048	0.004	2.718	20

The table1 summarizes the performance results of the EM model evaluated for horizon time of 20 minutes. The values of NRMSE and NMAE are equal to 0.048 and 2.718 respectively: this means that the model is ideal. The NMBE values for the EM algorithm-based filter model are positive (0.004) for the 20-minute horizon. The model tends to overestimate the values measured at the study site.

Figure 2 shows the correlation between the measurements and the prediction obtained with the EM algorithm for 20-minute time bin. It is possible to note most of the data points are concentrated around the straight line of the linear function. This graph tells us that there's a good agreement (R=0.99) between the predicted data and the measurements on the study site.

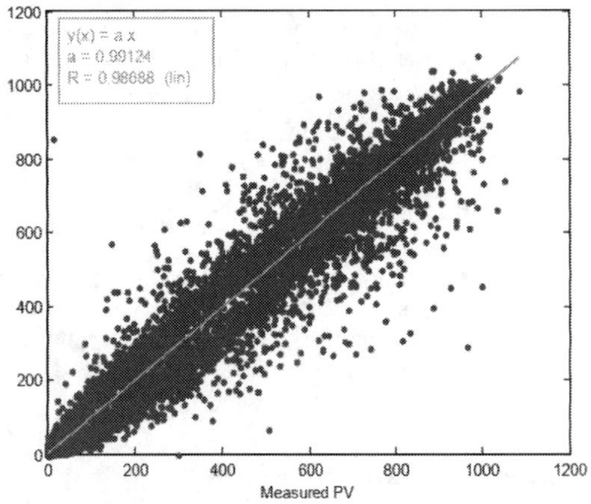

Figure 2. Correlation between the measurements made on the study site and those predicted by the filter model based on the EM algorithm.

Figure 3 presents the evolution of the values observed on the study site (yellow color), values predicted by the filter model based on the EM algorithm taking into account the impact of temperature and humidity (curve in green) and those predicted by the same model ignoring these atmospheric parameters (curve in blue). The filter model with and without impact of the atmospheric parameters (temperature and relative humidity) follows well the measured data.

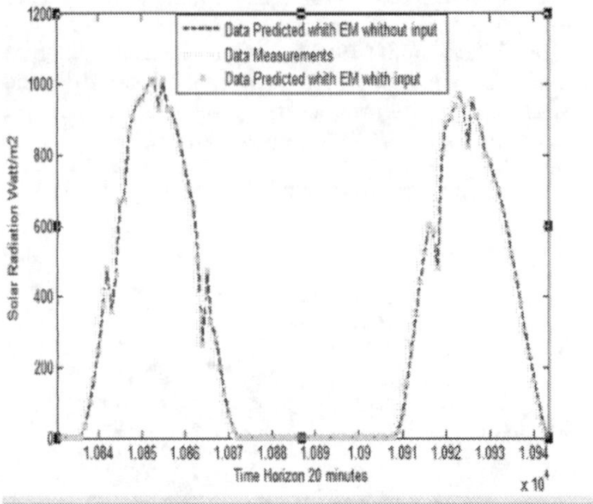

Figure 3. Evolution of the data measured and predicted by the Kalman filter via EM with input and without input.7

4 Conclusion

As part of this work we applied the kalman filter model to measured data on the Dakar (Senegal) site in order to make a short-term prediction of solar potential. The prediction is based on the Kalman filter model combined with the EM algorithm for estimating the parameters needed to execute the filter. We evaluated the model on the time horizon 20 minutes. The precision of this technique is studied with a set of standardized performance criteria taking into account the square root of the mean squared error (NRMSE), the absolute mean of the error (NMAE), the mean error of bias (NMBE). The simulation results following the 20-minute horizon show a good correlation between the prediction and the measurement with an NRMSE of 4.8%, an NMAE of 0.27% and an NMBE of 0.04%. This model can be used to predict the photovoltaic solar potential and the production of solar power plants for 20-minute time horizon on the DAKAR site in Senegal. It could contribute to the planning of energy providers based on photovoltaic power plant in Sahelo-Sahelian environment.

Acknowledgments

I am very grateful to the center of excellence in mathematics, computer science and ICT (CEA-MITIC) of Gaston Berger University of Saint-Louis (Senegal) for funding research mobility for my participation in the 2018 3rd International Conference on Sustainable and Renewable Energy Engineering in Zamora, Spain.
I thank the world science laboratory: c/o CERN,LAA Building # 29,CH-1211 Geneva 23, Switzerland, E-mail: info@worldlab.ch.

Thanks to the fellowships of the World Laboratory of Scientists during my 3 years of thesis: I was able to do my research quietly without problem.

https://doi.org/10.1051/e3sconf/20185701004

References

1. A. K. Yadav and S. S. Chandel. Solar radiation prediction using Artificial Neural Network techniques. Sustainable Energy, 2014; 33:772-**781.**
2. M. Ikhsan and al. Study of Renewable Energy Sources Capacity and Loading Using Data Logger for Sizing of Solar-wind Hybrid Power System. Procedia Technol, 2013; 11:1048-**1053.**
3. A. Mellit and al. Short-term forecasting of power production in a largescale photovoltaic plant. Solar Energy, 2014; 105:**401-413.**
4. H. Madsen and al. A tool for predicting the wind power production of offshore wind plants.in Proceedings of the Copenhagen Offshore Wind Conference &Exhibition, 2005.
5. PJ. Brockwell and Davis RA. Time series: theory and methods. Springer series in statistics, second edition, 2006.
6. JD.Hamilton. Times series analysis, 1994.
7. T.Soubdhan, J Ndong and al. A robust forecasting framework based on the Kalman filtering approach with a twofold parameter tuning procedure: Application to solar and photovoltaic prediction. Solar Energy, 2016; 131:**246-259.**
8. DJ.Docimo and al. Extended Kalman Filtering to estimate temperature and irradiation for maximum power point tracking of a photovoltaïque. Energy, Elsevier, 2017; 120:47-**57.**
9. J.Antonanzas and al. Review of photovoltaic power forecasting. Solar Energy, 2016; 136: **78-111.**
10. G.Lorenzo and al. Day-Ahead Hourly Forecasting of Power Generation from Photovoltaic Plants.IEEE Transactions on Sustainable Energy, 2017;**99:1-1.**
11. M.Pierro and al. Data-drivenupscaling methods for regional photvoltaic power estimation and forecast using satellite and numerical weather prediction data. Solar Energy, 2017; 158:**1026-1038.**

E3S Web of Conferences 57, 02001 (2018)
ICSREE 2018
https://doi.org/10.1051/e3sconf/20185702001

Responsiveness comparison between a lift-type and drag-type rotor in waves

Yingchen Yang[1], *Fredrick* Jenet[1], *Deyanira* Jimenez[2], *Brandon* Benavides[2], *Jesus* Cerda[2], and *Angel* Lopez[2]

[1]University of Texas Rio Grande Valley, Brownsville-Edinburg-Harlingen, TX, USA
[2]Instituto Tecnológico de Matamoros, Matamoros, Mexico

Abstract. Our recent progress on development of a vertical-axis unidirectional rotary wave energy converter (WEC) is discussed in this work. The WEC features a vertical-axis rotor that preforms unidirectional rotation in waves. The vertical axis arrangement makes the WEC respond well to waves from any direction with no realignment needs. And, the unidirectional behavior of the rotor promises no wave-frequency discrimination, which is in comparison to reciprocating WECs that employ the resonant principle and are very frequency-specific. In our earlier proof-of-concept studies, we have successfully demonstrated two types of rotor designs: a lift type employing hydrofoil blades and a drag type using cup blades. In the present work, the two rotor types were further explored experimentally by employing more rotor configurations and blade shapes. The focus was on revealing the rotor responsiveness in simulated waves under a freewheeling condition. The experimental results were compared between a lift-type and drag-type rotor. The comparison provided in-depth understanding on common features of the two rotor types and major differences between them. The yielded research findings will directly guide the development of a prototype vertical-axis unidirectional WEC.

1 Introduction

In a huge variety of wave energy converter (WEC) designs, a wave-driven reciprocating motion of a WEC (e.g., heaving up and down, pitching/rocking back and forth, bending/curling on and on, etc.) has been a main theme [1]. Such WEC designs take the advantage of the resonant principle to gain a high efficiency in a favorable range of the wave frequency. For power take-off, many reciprocating WEC designs employ either hydraulic circuits (e.g., Pelamis Wave Power) or other mechanical means [2,3] to convert the WEC's primary reciprocating motion into a secondary unidirectional rotation to be transmitted to a traditional electric generator. Some other WEC designs use a direct electric generator to avoid this type of conversion [4].

Differing from the reciprocating WECs, unidirectional WECs feature unidirectional rotation directly driven by waves [5,6]. This feature makes unidirectional WECs fit well in a broad range of the wave frequency since they do not rely on the resonant principle. The unidirectional rotation could be either about a horizontal axis [5] or about a vertical axis [6]. A horizontal-axis unidirectional WEC demands to have its axis approximately aligned along the wave crest direction. Therefore, if the wave direction changes, WEC realignment is needed. In contrast, a vertical-axis unidirectional WEC is insensitive to the wave direction and, thus, has no needs for realignment.

The undoubted tolerance to the wave frequency and wave propagation direction makes the vertical-axis WEC class very attractive. A successful design of this type would demonstrate a great potential towards a simple and robust WEC for low-cost electricity. Due to the complex flow condition resulted by waves, however, it is a great challenge to realize and optimize such a WEC design. In waves, water particles perform orbital motion with drift [7,8]. Therefore, wave-driven water flow is unsteady and omnidirectional. In pursuit of a vertical-axis unidirectional rotor, the omnidirectional water flow is expected to drive the rotor for unidirectional rotation. From the hydrodynamics perspective, it would be ideal if a local flow in any spatial direction could always produce a strong torque along a predefined direction of the vertical axis to drive the rotor for unidirectional rotation. This way, the rotor would be capable of unconditionally absorbing energy from waves in a relatively consistent rather than an intermittent way.

Recently the authors have conducted proof-of-concept studies on a few vertical-axis WEC designs [9,10]. The present work focuses on characterization and comparison of two typical rotor types—a drag type and a lift type—in a freewheeling mode by experimental means. For each rotor type, various blade shapes are explored and the formed rotors are examined in various periodic and directional flows. The main objective is to quantify the unidirectionality of the rotors' angular velocity in response to the above flows in terms of the averaged values, time traces, and amplitude spectra.

2 Experimental system

The experiments were carried out in a wave flume of inner dimensions 15m (L) × 1m (W) × 1.3m (H). To characterize a rotor's unidirectional responsiveness to omnidirectional flows, a servo-motor-driven translation carriage developed in house[11] was employed to translate the rotor in quiescent water in the flume. The carriage can translate the rotor in three modes while maintaining the rotor axis/shaft vertical all the time and allowing the rotor to rotate freely about the shaft. The three modes include horizontal oscillation, vertical oscillation, and circular motion in a vertical plane. The circular motion of the rotor in quiescent water approximately simulates a stationary rotor (with one degree of freedom about the shaft) in deep waves where the water particles perform a circular motion (with Stokes drift neglected) [7]. The horizontal and vertical oscillations represent two components of the circular motion. These two component flow conditions are intended for examining the rotor's responsiveness to flows along the two directions. Using the translation carriage to move the rotor in quiescent water makes it much easier to create relative flows along different directions and vary wave parameters than exposing the rotor in waves. For this reason, no real waves were employed in this study. The frequency and peak-to-peak amplitude associated with the three motion modes are herein defined as the simulated wave frequency f_W and simulated wave height H. The experimental system with a drag-type rotor on is depicted in figure 1.

Fig. 1. A photo of the experimental system.

Two types of rotors were examined and compared. They were a drag type and a lift type, (figure 2). A drag-type rotor (figure 2a) employs cup-shaped blades, whereas a lift-type rotor uses wing-shaped blades (figure 2b); all the blades were 3D printed, polished, and painted. To make the two rotor types comparable, some common constraints were applied to all the rotor designs. First, each type consists of two sets of triple blades. Within a set, the three blades show an even circumferential distribution. Between the two sets, there is an angular offset about the shaft and a distance along the shaft to be 60° and 152.4 mm, respectively. Secondly, both types have the same rotor diameter D = 500 mm. The rotor diameter D is the diameter of the maximum circular swept area of the blades when rotating about the shaft.

For each rotor type, blades with various

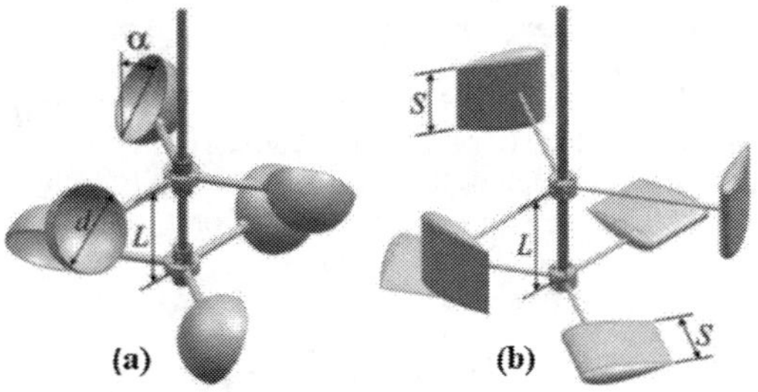

Fig. 2. A drag-type rotor (a) and a lift-type rotor (b).

hydrodynamic shapes were considered. Specifically, in the drag-type rotor design, two cup shapes were employed as blades (figure 3). One was a hemispherical cup and the other a NACA0035-nose cup. Both cups had the same open-end outer diameter of d = 127 mm and the same wall thickness of 1.5 mm. The NACA0035-nose cup was formed by revolving a NACA0035 profile of a chord length 363 mm about the chord line and trimming at a chord-wise distance of 108.9 mm from the nose. Four drag-type rotors were configured using these two types of cup blades: Semi-0° (six hemispherical cups with no deflection angle, α = 0°), Semi-30° (six hemispherical cups with a deflection angle, α = 30°, figure 2a), NACA-0° (six NACA0035-nose cups with no deflection angle, α = 0°), and NACA-30° (six NACA0035-nose cups with a deflection angle, α = 30°, similar to figure 2a).

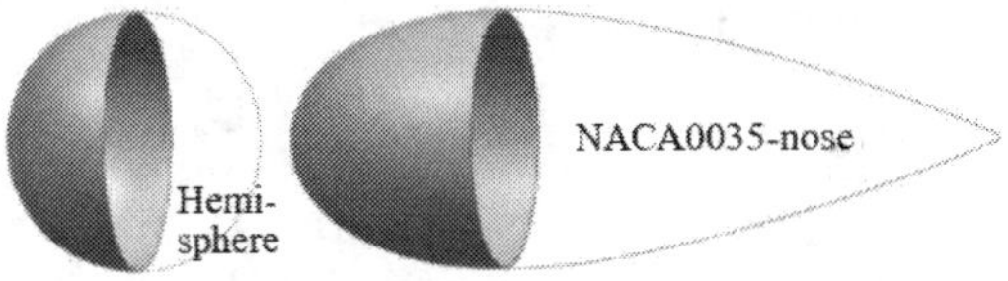

Fig. 3. Two cup shapes for drag-type rotors.

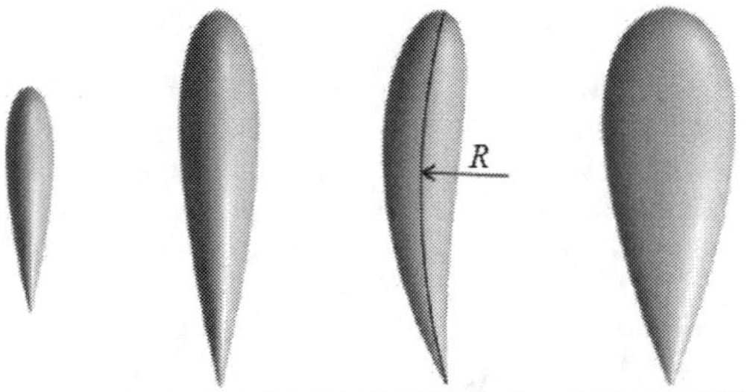

Fig. 4. Four hydrofoil profiles for lift-type rotors.

In the lift-type rotor design, four hydrofoil profiles as specified in figure 4 were adopted. The NACA0021-cambered profile had a circular camber line of radius R = 236.7 mm. By extruding each of the four 2-D profiles along the third dimension and rounding both ends to form a wing-like blade, all the four blade shapes had the same span of S = 101.6 mm (excluding the rounded ends, figure 2b). Four lift-type rotors were configured using these four blade shapes: NACA0021-small (six NACA0021 blades with a chord length of C = 76 mm), NACA0021 (six NACA0021 blades with C = 127 mm, figure 2b), NACA0021-cambered (three horizontal NACA0021 blades and three vertical NACA0021-cambered blades, all with C = 127 mm), and NACA0035 (six NACA0035 blades with C = 127 mm).

By mounting a rotor to the aforementioned translation carriage through a shaft-bearing-housing mechanism, the rotor can freely rotate about its shaft with low mechanical friction. If there is no power take-off from the rotor, such a rotation condition is termed as freewheeling. All the four drag-type rotors and four lift-type rotors were tested in a freewheeling mode. And, they were all tested in the same range of the simulated wave frequency $0.33 \le f_W \le 0.83$ at a fixed wave height H/D = 0.635. For each experimental run, a 75-second video was taken to record a rotor's real-time angular displacement. A dial that rotated together with the rotor was employed for this purpose. The video was then processed to derive the time trace of the rotor's angular velocity. Details on data acquisition and post processing to obtain time traces were discussed in our earlier work[11]. To determine an averaged angular velocity and standard deviation, three repetitive runs were conducted for each experimental condition considered. The standard deviation turned out to be very low due to the relatively long recording time of individual runs. Therefore, no error bars were provided when presenting the averaged angular velocity.

3 Results and discussion

Aiming at characterizing vertical-axis rotors' unidirectional behavior in omnidirectional flows and comparing the two rotor types, experimental results are sorted by the rotor type first and a comparative discussion is provided towards the end.

3.1 Drag-type rotors

For the four drag-type rotors (Hemi-0°, Hemi-30°, NACA-0°, and NACA-30°) in three types of the carrying motion (circular motion, horizontal oscillation, and vertical oscillation), the averaged angular velocity $\overline{\omega}$ versus the simulated wave frequency f_W are presented in figure 5. In circular carrying motion, figure 5a shows an approximately linear trend of $\overline{\omega} \sim f_W$ for each rotor, and no significant difference among the four rotors is observed. A couple of noticeable differences are still evident though: (i) at a fixed deflection angle (0° or 30°), a rotor with hemispherical blades has slightly higher $\overline{\omega}$ than a rotor with NACA-nose blades and, (ii) with a selected blade shape (hemisphere or NACA nose), a rotor with a 30° deflection angle resulted in a somewhat higher $\overline{\omega}$ than a rotor with a 0° deflection angle.

It is known from cup anemometer study [12] that a cup rotor with a 0° deflection angle responds well with flows normal to the rotor shaft. The employment of a 30° angle is intended to enhance the rotor responsiveness to flows along the shaft direction [9]. Figure 5a indirectly supports this intention. More insight is gained from figure 5c for rotors in vertical oscillatory flows. Specifically, between Hemi-0° and Hemi-30° the deflection only leads to a minor increase in $\overline{\omega}$ in relatively high values of f_W. Between NACA-0° and NACA-30°, however, a large increase has been observed, and it gets stronger with the increasing f_W. It is noteworthy though, that such enhancement to vertical flows is achieved by sacrificing the rotor responsiveness to horizontal flows, as evident in figure 5b. It is particular severe on the rotor NACA-30°; it demonstrates the strongest rotation among the four rotors in vertical oscillatory flows (figure 5c) but the weakest in horizontal oscillatory flows (figure 5b). Further

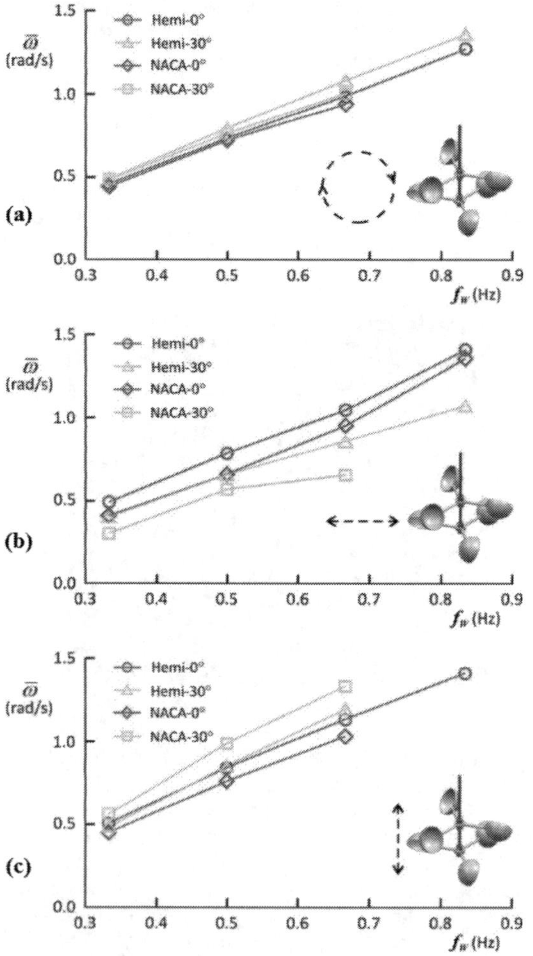

Fig. 5. Averaged angular velocity $\overline{\omega}$ versus the simulated wave frequency f_W for four drag-type rotors in three carrying motion types. (a) Circular motion. (b) Horizontal oscillation. (c) Vertical oscillation.

comparison between figures 5b and 5c leads to another very interesting observation: both the rotors with no deflection (i.e., Hemi-0° and NACA-0°) already rotate as well in vertical oscillatory flows as in horizontal oscillatory flows.

Time traces and associated spectra reveal further details on the rotors' unidirectional behavior in response to flows in different directions. Figures 6 through 9 present such results for each of the four drag-type rotors at a selected wave frequency $f_W = 0.67$ Hz. Time t in the time traces is normalized by the wave period T, $T = 1/f_W$, and the frequency f in the spectra is normalized by f_W. Both common features and distinctions are evident as presented by these figures.

With a specified rotor and among the three types of carrying motion, the differences among the time traces are easy to tell in terms of the waveforms and levels of fluctuation (e.g., figures 6a, 6c, and 6e). Such differences are also reflected in the corresponding spectra with different numbers and levels of peaks (e.g., figures 6b, 6d, and 6f). With a specified type of carrying motion and among the four rotors, however, waveform resemblance among the time traces is evident (e.g., figures 6a, 7a, 8a, and 9a) except for the vertical carrying motion. Obviously, in the vertical carrying motion a rotor responsiveness is largely affected by the blade shape and deflection angle (figures 6e, 7e, 8e, and 9e). For all the four rotors and three carrying motion types, an overly dominant peak always occurs at twice the simulated wave frequency in every spectrum. A noticeable peak at the wave frequency does exist in the circular carrying motion (figures 6b, 7b, 8b, and 9b), but is almost negligible in the horizontal and vertical carrying motion. In addition, there are one or two discernible peaks nearby the peak at the wave frequency in the circular carrying motion (figures 6b, 7b, 8b, and 9b), and the physics is unclear yet.

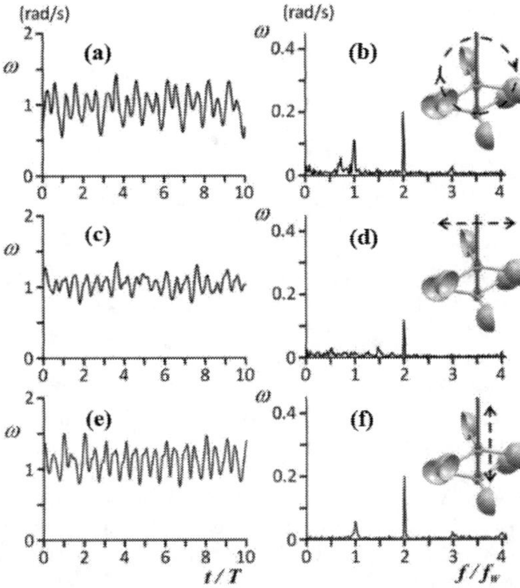

Fig. 6. Time traces (left column) and amplitude spectra (right column) of the Hemi-0° rotor's angular velocity in response to three carrying motions as specified by inserts on the right.

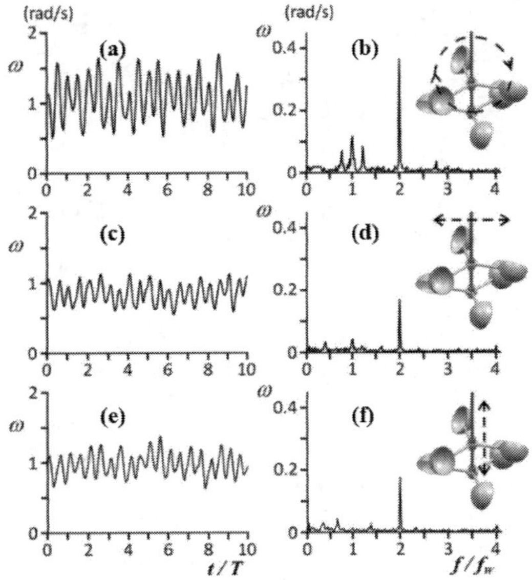

Fig. 7. Time traces (left column) and amplitude spectra (right column) of the Hemi-30° rotor's angular velocity in response to three carrying motions as specified by inserts on the right.

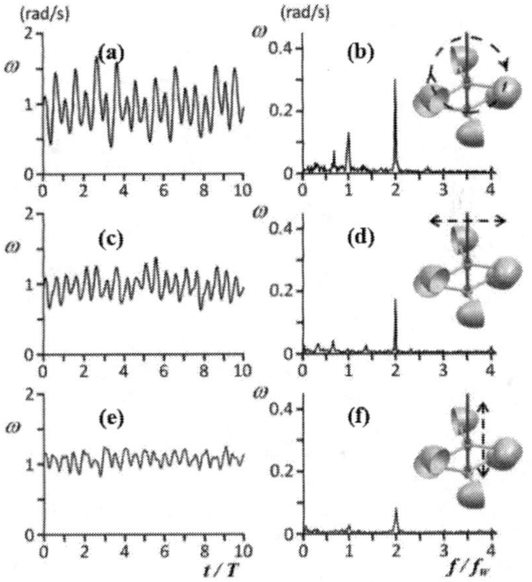

Fig. 8. Time traces (left column) and amplitude spectra (right column) of the NACA-0° rotor's angular velocity in response to three carrying motions as specified by inserts on the right.

3.2 Lift-type rotors

The averaged angular velocity $\overline{\omega}$ versus the simulated wave frequency f_W for the four lift-type rotors (NACA0021, NACA0021-cambered, NACA0021-small,

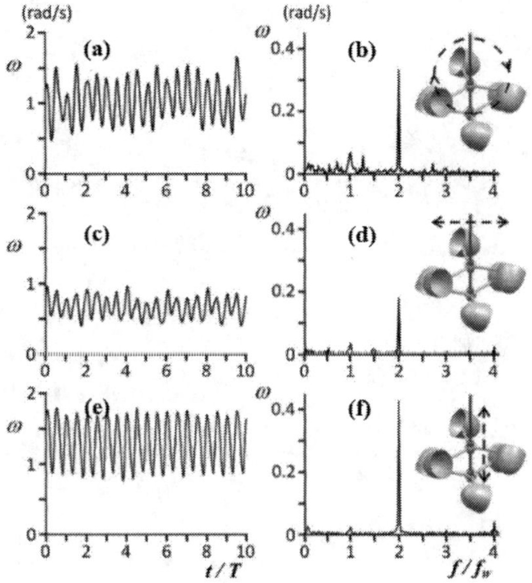

Fig. 9. Time traces (left column) and amplitude spectra (right column) of the NACA-30° rotor's angular velocity in response to three carrying motions as specified by inserts on the right.

and NACA0035) are presented in figure 10. Specifically, figures 10a, 10b, and 10c compare the four rotors in circular, horizontal, and vertical carrying motions, respectively. A general pattern in figure 10 is that $\overline{\omega}$ increases with f_W for each case considered. For different rotors in different carrying motions, however, individual trends are quite different. Another common behavior is that, the four rotors all demonstrate a lower $\overline{\omega}$ in the horizontal carrying motion than in the circular or vertical ones. Such a weak responsiveness to horizontal flows can be addressed by properly increasing the span of the three vertical blades in a rotor design.

Among the four rotor designs, comparing NACA0021 with NACA0021-small provides understanding on how the blade chord length (127 mm versus 76mm) changes a rotor's rotation. Similarly, comparing NACA0021 with NACA0035 helps quantify the effect of the blade thickness (26.7 mm versus 44.5 mm), and comparing NACA0021 with NACA0021-cambered allows determination of the blade cambering effect (straight versus cambered).

As shown in figure10, an increase in the chord length (from NACA0021-small to NACA 0021) leads to a quite noticeable increase in $\overline{\omega}$ in all the three carrying motions. In contrast, an increase in the blade thickness (from NACA0021 to NACA0035) results in a slight decrease in $\overline{\omega}$ in circular motion, a negligible change in horizontal oscillation, and a relatively large increase in vertical oscillation at lower values of f_W. The most remarkable difference is made by the blade cambering (from NACA0021 to NACA0021-cambered); a significant increase in $\overline{\omega}$ has been achieved in all the three carrying motions, and the increase enhances with increasing f_W.

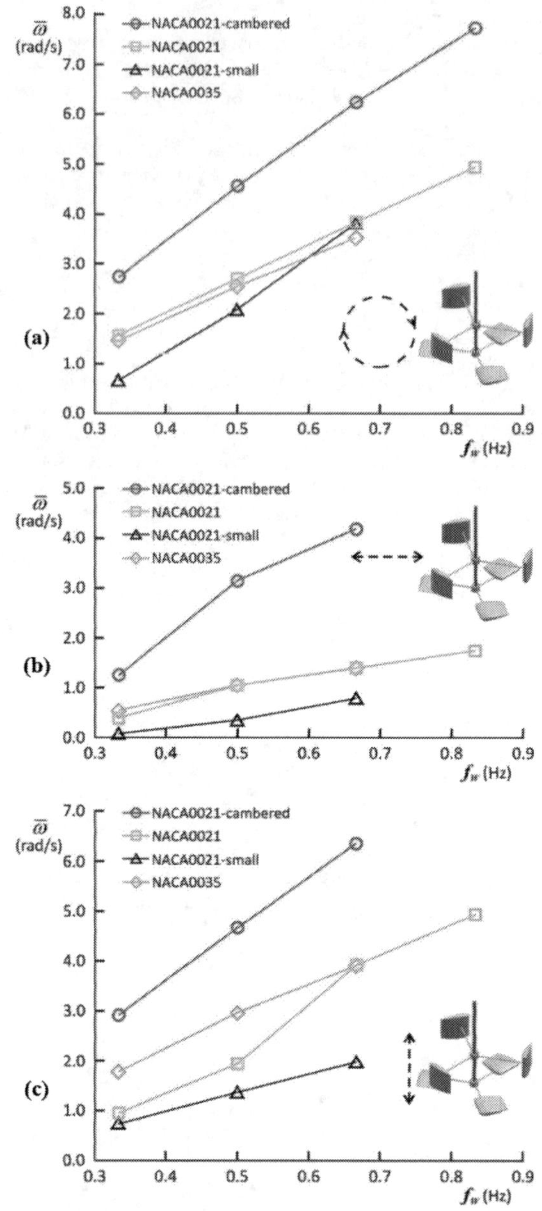

Fig. 10. Averaged angular velocity $\overline{\omega}$ versus the simulated wave frequency f_W for four lift-type rotors in three carrying motion types. (a) Circular motion. (b) Horizontal oscillation. (c) Vertical oscillation.

Overall, the NACA0021-cambered rotor demonstrates the best performance among the four rotors.

Time traces and spectra for the four lift-type rotors in three carrying motions and at a selected wave frequency f_W = 0.67 Hz are shown in figures 11 through 14. Note that all the time traces are plotted in the same scale but in various ranges of ω.

Similar to the drag type-rotors, the lift-type rotors yield time traces in different waveforms and fluctuation levels in response to different carrying motions. For

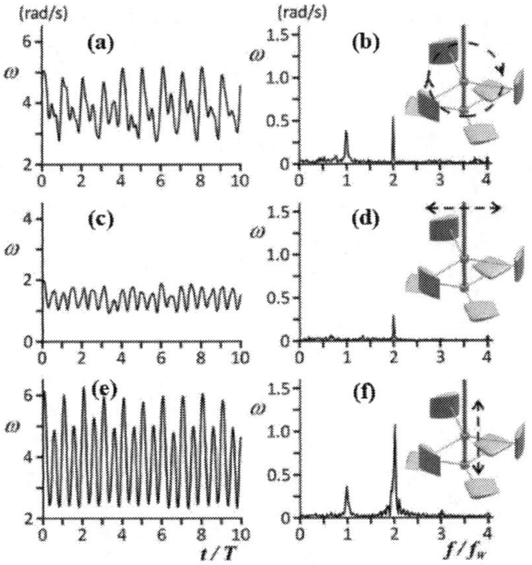

Fig. 11. Time traces (left column) and amplitude spectra (right column) of the NACA0021 rotor's angular velocity in response to three carrying motions as specified by three inserts on the right.

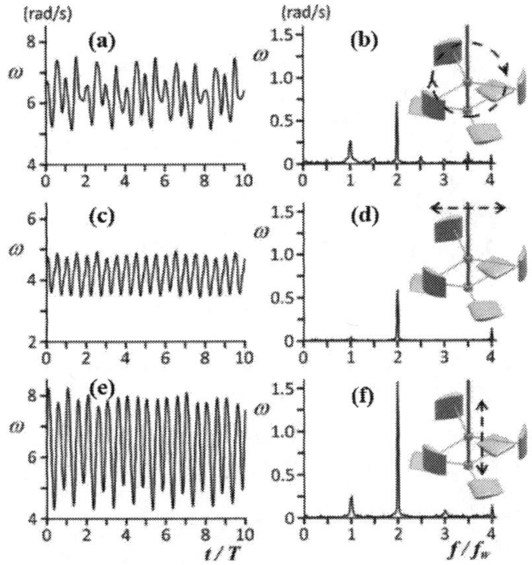

Fig. 12. Time traces (left column) and amplitude spectra (right column) of the NACA0021-cambered rotor's angular velocity in three carrying motions as specified by three inserts on the right.

example, in the vertical carrying motion, both the rotors NACA0021 and NACA0021-cambered gain a strong rotation in terms of the mean value and fluctuation of ω (figures 11e and 12e). In the horizontal carrying motion, however, the mean value and fluctuation drops largely (figures 11c and 12c). Comparison between figures 11 and 12 shows that, both the horizontal and vertical carrying motions result in time traces close to a rectified waveform with an offset for the NACA0021 rotor

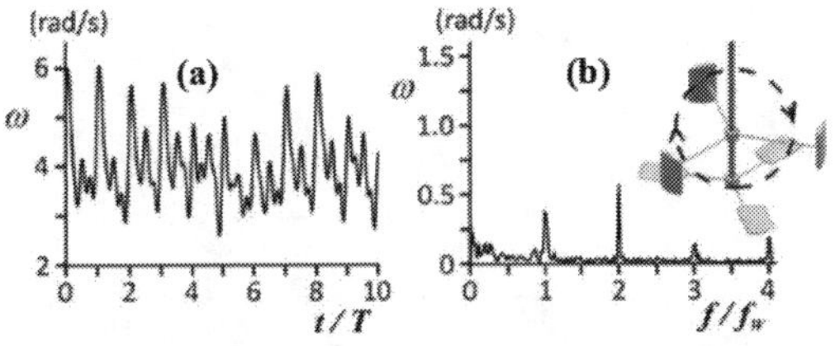

Fig. 13. Time traces (left) and amplitude spectra (right) of the NACA0021-small rotor's angular velocity in response to the circular carrying motion as specified by the insert on the right.

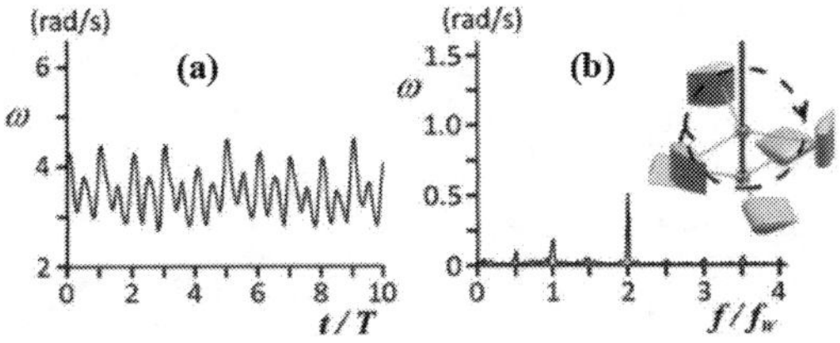

Fig. 14. Time traces (left) and amplitude spectra (right) of the NACA0035 rotor's angular velocity in response to the circular carrying motion as specified by the insert on the right.

(figures 11c and 11e), but it is not the case for the NACA0021-cambered (figures 12c and 12e). The waveforms in the circular motion for the lift-type rotors are quite similar (figures 11a, 12a, and 14a). They also resemble that for the drag-type rotors (figures 6a, 7a, 8a, and 9a) to a large extent. An exception is the NACA0021-small rotor in the circular carrying motion; the time trace (figure 13a) is much noisier than others. The corresponding spectrum (figure 13b) shows a dominant peak at $f/f_W = 2$ and a less dominant peak at $f/f_W = 1$. The noise is due to the discernable higher harmonics at $f/f_W = 3$ and 4, as well as some very low frequency interference.

The spectra in figures 11 through 14 all present a dominant peak at $f/f_W = 2$, which is, once again, similar to that for drag-type rotors. In contrast, the secondary peak at $f/f_W = 1$ is comparatively small in circular motion (figures 11a, 12a, 13a, and 14a), insignificant in vertical oscillation, and hardly discernable in horizontal oscillation.

3.3 Comparison

Using the averaged angular velocity $\overline{\omega}$ obtained from the circular carrying motion as a criterion, the best performing rotor from each of the two rotor types is selected. One is Hemi-30° from the drag type and the other is NACA0021-cambered from the lift-type. The two representative rotors are compared in terms of $\overline{\omega}$, as shown in figure 15. Obviously, under the same flow conditions the lift-type rotor rotates much faster than the drag-type rotor. Specifically, in the circular carrying motion the lift-type rotor rotates about 5.7 times as fast as the drag-type rotor in the whole testing range of f_W.

It has long been known that, in uniform flows, a lift-type rotor is generally more efficient than a drag-type

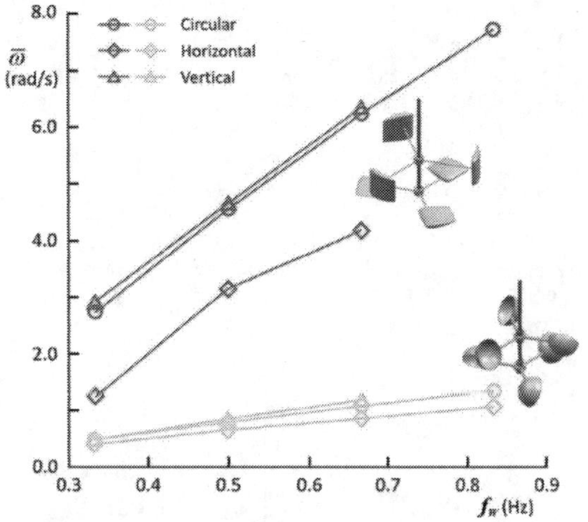

Fig. 15. Comparison of $\overline{\omega}$ between a drag-type rotor (Hemi-30°) and a lift-type rotor (NACA0021-cambered) among three motion types.

rotor. The much stronger flow separation in the drag-type rotor results in more loss of the flow's hydrokinetic energy on the viscous dissipation. In wave-rotor interaction, however, this common understanding may need to be re-evaluated. Due to the omnidirectional nature of water flow in waves, an instantaneous angle of attack of a local flow with respect to a lift-type rotor blade could sometimes go very large (e.g., 90° or beyond), leading to a flow separation as severe as, or even worse than, the drag-type case.

While figure 15 still supports the common understanding, the data were obtained in a freewheeling mode. In this case the lift-type rotor rotates very fast. Hence it might be able to keep the angles of attack of individual blades in waves in a relatively small level. In a power take-off mode, however, the rotor's rotation may need to be reduced to a lower level due to a variety of considerations (e.g., an added resistive load, cavitation avoidance, power output maximization, etc.). Then the slowed rotation may lead to a large increase in the angles of attack and make the rotor loose its advantages. Therefore, a more comprehensive comparison between the two rotor types by employing a power take-off mechanism needs to be considered in future research.

4 Conclusions

Two types of vertical-axis unidirectional rotors—a drag type and a lift type—were experimentally explored in simulated waves in a freewheeling mode. In each rotor type, four rotors with different blade shapes and dimensions were employed. The rotors' unidirectional responsiveness was characterized in terms of the time traces, spectra, and averaged values of a rotor's angular velocity. In simulated waves, all the rotors demonstrated highly unidirectional rotation, yet with fluctuation. An overly dominant frequency of the fluctuation occurred at twice the simulated wave frequency for all the rotors.

Two selected rotors, with each representing one rotor type, were compared by means of the averaged angular velocity. The lift-type rotor rotated 5.7 times as fast as the drag-type rotor.

References

1. M.S. Guney, K. Kaygusuz, *Renew. Sust. Energ. Rev.* **14**, 2996-3004 (2010)
2. J Ai, H. Lee, C. Liang, L. Zuo, *Proc. ASME 2014 Int. Design Engineering Technical Conf.* V008T11A086 (2014)
3. A. Battezzato, *J. Theor. App. Mech.*, **53**, 195-207 (2015)
4. M. Leijon, H. Bernhoff, O. Agren, J. Isberg, J. Sundberg, et al., *IEEE Trans. Energy Convers.* **20**, 219–24 (2005)
5. M.R. Ahmed, M. Faizal, Y.H. Lee, *Ocean Eng.* **65**, 32-8 (2013)
6. E.A. Rossen, P.C. Scheijgrond, R. Mikkelsen, *Proc. 4th European wave Energy Conf.* 2389448844 (2000)
7. G.L. Pickard, S. Pond *Introductory Dynamical Oceanography* 2nd ed (Butterworth-Heinemann, 1983)
8. O.M. Phillips, *The dynamics of the upper ocean* 2nd ed (Cambridge University Press, 1977)
9. Y. Yang, I. Diaz, M. Morales, P. Obregon, *Proc. ASME 34th Int. Conf. on Ocean, Offshore and Arctic Engineering* doi:10.1115/OMAE2015-41140 (2015)
10. Y. Yang, I. Diaz, S. Soto, *Proc. ASME 2016 Int. Mechanical Engineering Congress & Exposition* doi:10.1115/IMECE2016-65952 (2016)
11. Y. Yang, F. Salazar, J. Soto, *Proc. ASME 2017 Fluids Engineering Summer Conf.* doi:10.1115/FEDSM2017-69458 (2017)
12. S. Pindado, J. Cubas, A. Sanz-Andrés, *Sci. World J.* 197325 (2013)

Energy optimization of chillers by automating a Cooling System

Wilson E. Sánchez[1], *Bryan L.* Robayo[1], *Pablo E.* Rodríguez[1], *Martha E.* Salazar[2], *Santiago* Jácome[1] and *Alvaro S.* Mullo[1]

[1]Departamento de Eléctrica y Electrónica, Universidad de las Fuerzas Armadas ESPE, ID: 60104598, Sangolquí, Ecuador
[2]Departamento de Ciencias Exactas, Universidad de las Fuerzas Armadas ESPE, ID: 60104598, Sangolquí, Ecuador

Abstract. This publication presents the work done on energy optimization of a group of chillers by automating the cooling system of blowing dies in an industry that is entrusted with the manufacture of plastic containers. In a first stage, the preliminary study on the energy analysis of the old system, where each chiller was in charge of cooling a specific group of blow moulding machines, without taking into account whether or not the blow moulding machines are working, is detailed; subsequent to this, the implementation of the automated cooling system was carried out, where all chillers were centralized in a machine room, automation was carried out with a Siemens S7-1200 PLC, switching the chillers on and off is based on the total thermal capacity required by the blow moulding machines. Finally, the energy analysis of the new cooling system was carried out, where the results were compared, and conclusions were drawn.

1 Introduction

Recent reports and perspectives on electricity show a widening gap between electricity generation and electricity demand. Energy costs are therefore a determining factor in economic and social prosperity. As a result, policy, industry and research focus both on the exploitation of alternative energy sources for energy generation and on the development of concepts, methodologies and technologies to reduce energy consumption. An outstanding challenge is energy analysis and optimization of complex systems, such as industrial automation systems. The individuality and dynamics of the system environment during operation make it difficult to identify goal-oriented optimization measures. This contribution proposes a model-based and user-centered approach to automatic energy analysis. This approach helps users to reduce their energy costs through automated generation and implementation of individual proposals for optimizing electrical energy for their industrial automation systems [1].

The availability of conventional energy resources is declining [2]. At the same time, the demand for electrical energy is increasing due to the constant increase in the number of devices and systems that consume electrical energy [3]. These conflicting trends lead to an increase in energy costs affecting all economic and social sectors. The efforts of politics, industry and academia are aimed at both alternatives for power generation and technologies and concepts to reduce the energy consumption of electrical devices and systems. Industrial automation systems are typical devices that consume electrical energy. These must be optimized to help meet the challenges ahead in the energy sector. Industrial automation systems comprise systems that contain a technical process within a technical system. The technical system consists of the communication system, computers, sensors, actors and process operators necessary to automate the technical process [4]. Technical systems are technical products or technical installations. Technical systems that can be characterized as mass products with few sensors and actuators and a high degree of automation are technical products (e. g. a washing machine). Technical plants are unique systems with a high number of sensors and actuators and a medium to high degree of automation. Industrial automation systems are widely used in private homes, workplaces or intersected areas such as public and commercial transport. Users must handle different and complex industrial automation systems. The growing number and increasing complexity of the systems capture much of the attention of users. Additional tasks, such as energy cost analysis, can be easily overloaded [5].

Today, the price of energy is rising. Awareness of the environmental sustainability of governments and customers has been increasing. As a result, manufacturing companies are increasingly motivated to reduce the energy cost of their production activities [6].

Many manufacturing processes require the careful control of surface temperatures, and heat transfer rates, to increase production and improve product quality. Injection moulding, blow moulding, die casting and extrusion are all examples of manufacturing processes that can benefit from incorporating methods of increased and balanced heat transfer within the tooling. For injection moulding, heat transfer is usually carried out using straight cooling channels drilled into the mould. As cooling accounts for a large portion of the total cycle time, this can lead to significant cost savings. Schmidt et al. [7] showed in one case study that 3D printed conformal cooling tool inserts reduced cooling and cycle times by 19-20% over conventionally machined inserts. Many other manufacturing processes may also benefit from advanced thermal management of the tooling. Shayfull et al. [8] concluded that rapid heat cycle moulding, in which the mould is heated and cooled for each cycle is also likely to benefit greatly from conformal cooling, however this is yet to be fully explored. Au and Yu carried out computer aided engineering simulations to demonstrate the benefits of conformal cooling over conventional cooling for blow moulding tooling [9]. Hölker et al. showed that conformal cooling of extrusion dies could reduce surface defects on the extruded material [10]. Armillotta et al. showed that conformal cooling improves the surface quality of die cast parts, reduces cycle time and reduces part porosity [11].

Improving the energy efficiency of manufacturing processes is one of the top priorities in the sustainable energy hierarchy and is a widely adopted strategy by industry to reduce energy usage, emissions and costs [12].

2 Experimental Investigation

The energy consumption of chillers is associated with the plant's production capacity. The start-up of the blowers generates a thermal load, which must be extracted by the chillers (Table 1). At this time, the blow moulding machines associated with the corresponding chillers worked at their maximum production capacity. The equipment used was the FLUKE 1735 Power Logger.

When a chiller starts up, it scans the system (pressure and coolant level, water level in the storage tank, temperature range setting), and if there are no alarms, it starts the main pump, which circulates the water through the system. Depending on the temperature of the water in the chiller tank, the compressor of the refrigeration circuit will turn on and by means of the coil located inside the chiller tank, cool the process water.

Table 1. Working Temperatures of the mould in operation.

Zone	Temperature [°C]
Extruded eyelid outer face	170
Outer face of extruded container	60
Outer side of matrix in contact with extruded parison	40
Water temperature at matrix inlet	7

2.1 Analysis of energy consumption chiller HENN M. T. A. TAE 101

Data collection with the FLUKE 1735 Power Logger analyzer started at 16h36 on December 1, 2016 and ended at 14h00 on December 2, 2016. In order to analyze in greater detail, the power values demanded by the chiller, a random hour lapse was taken, since, as can be seen in the general graph (figure 1), the power curve demanded is periodic, in addition it is observed that there are two levels of power consumed, one representing the exclusive consumption of the pump, and another higher power level, in which the consumption corresponds to the pump and the refrigeration compressor, working simultaneously.

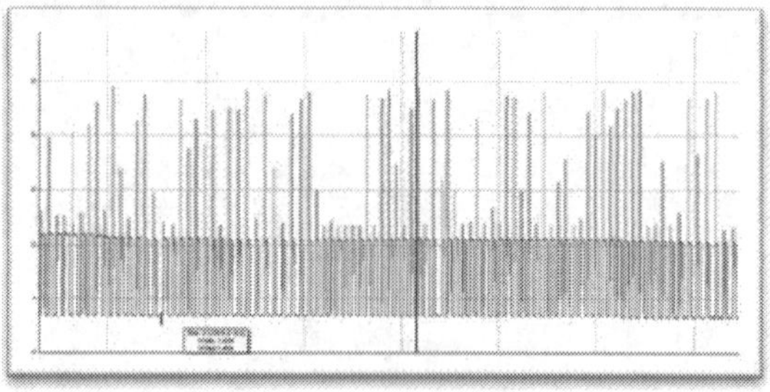

Fig. 1. Wer demanded by chiller HENN.

Figure 2. It is considered a one-hour interval (17:00 to 18:00), in which two different power values are distinguished in different periods of time.

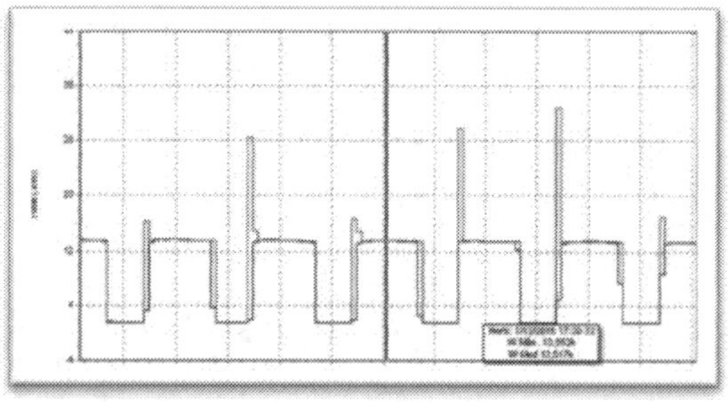

Fig. 2. Power required by the pump and compressor.

The highest value (13.517[kW]) corresponds to the combined power requirement between the main pump and the refrigeration circuit (compressor). It can be observed that the average duration of this power demand value is six minutes. The peak values shown in the graph correspond to the values of the compressor starting current, whose value is 32.84[kW].

The lowest required power value corresponds to the load of the pump in the hydraulic circuit, whose value is 1.505[kW], with an average duration of 4 minutes. (Figure 3).

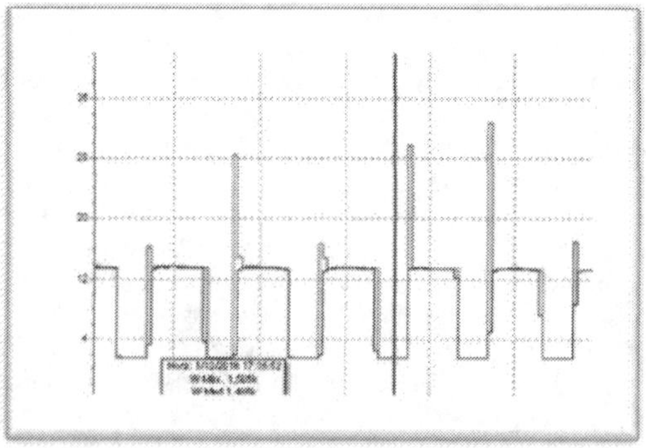

Fig. 3. Power required by the pump.

Analyzing figure 3, the duration of each cycle is 10 minutes, having six cycles in one hour, of which 60% of the time (0.6 h) the compressor and pump are active, while the remaining 40% only works the pump (0.4 h). With this data, the estimated energy consumption corresponding to the compressor of the HENN chiller is calculated. (E_{CHENN})

$$E_{CHENN} = Time \times Power \quad (1)$$

$$E_{CHENN} = 0.6\,[h] \times 13{,}517\,[KW]$$
$$E_{CHENN} = 8.11\,[KWh]$$

Similarly, the power required by the HENN chiller pump is calculated. (E_{BHENN})

$$E_{BHENN} = Time \times Power$$
$$E_{CHENN} = 0.4\,[h] \times 1{,}505\,[KW]$$
$$E_{CHENN} = 0.602\,[KWh]$$

The total energy consumed of the chiller HENN ($E_{HENNcalc}$), in one hour will be then:

$$E_{HENNcalc} = E_{CHENN} + E_{BHENN} \quad (2)$$

$$E_{HENNcalc} = 8.1\,[KWh] + 0.6\,[KWh]$$
$$E_{HENNcalc} = 8.713\,[KWh]$$

To check the data, a comparison is made with the values obtained in the energy analyzer at the same time interval. (Figure 4)

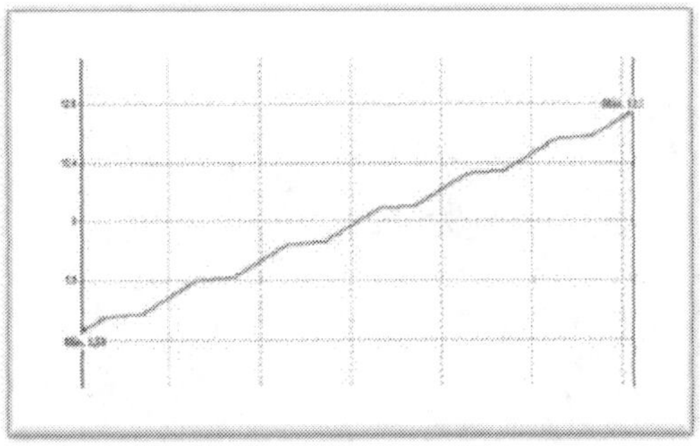

Fig. 4. Energy consumption in one hour of HENN chiller.

$$E_{HENNdat} = E_{18H00} - E_{17H00}\ [kWh] \quad (3)$$

$$E_{HENNdat} = 12{,}324 - 3{,}58\,[kWh]$$
$$E_{HENNdat} = 8{,}74\,[kWh]$$

Comparing the calculated energy value ($E_{HENNcalc}$) with the energy value obtained by the Power Logger ($E_{HENNdat}$) of the HENN branded chiller confirms that the calculated values are correct. This energy value is taken as a basis for the calculation of energy consumed monthly by the chiller days.

$$E_{HENNmes} = E_{HENNcal} * 24[h] * 30[dias] \quad (4)$$

$$E_{HENNmes} = 8{,}713[kWh] * 24[h] * 30[dias]$$
$$E_{HENNmes} = 6272.78\ [kWh/mes]$$

2.2 Energy consumption analysis KOOLANT KOOLERS HCV-7500

Data collection was done with the FLUKE 1735 Power Logger analyzer, started at 15h12 on 30 November 2016, and ended at 15h34 on 1 December 2016.

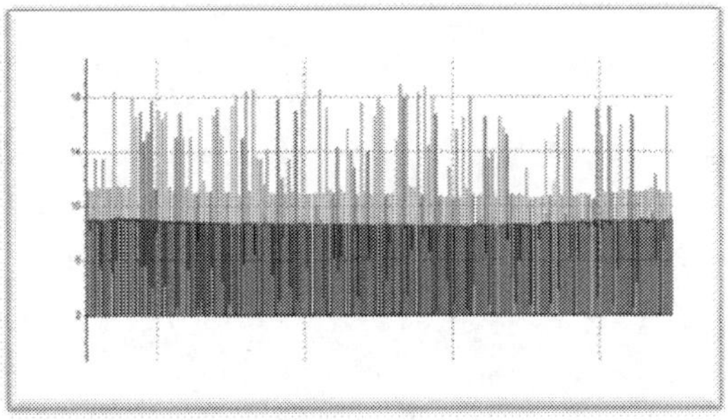

Fig. 5. Power required by the chiller KOOLANT.

To analyze in greater detail, the power values demanded by the chiller, a random hour lapse was taken, since as can be seen in the general graph (figure 5), and the power curve demanded is periodic.

In figure 6, it is considered a one-hour interval (17:00 to 18:00), in which two different power values are distinguished in different periods of time. The power value for the pump starts at 17:26:44 hours (figure 6) and ends at 17:29:44 hours (figure 7), thus determining a time of three minutes, with an average power value of 1.943[kW].

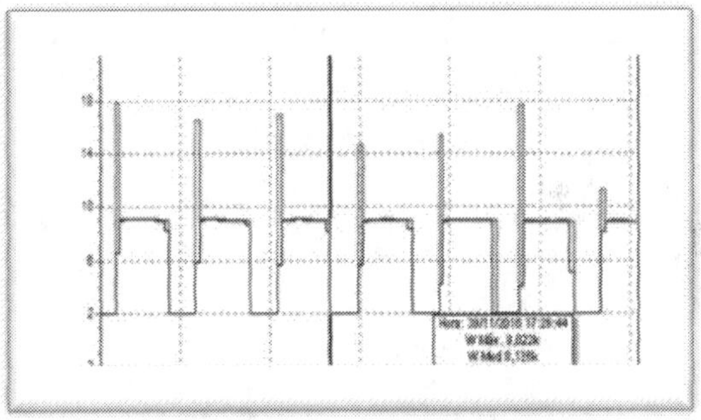

Fig. 6. Start pump working time.

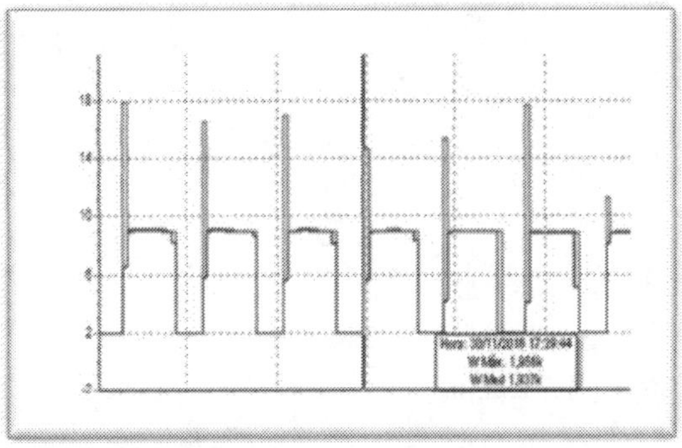

Fig. 7. End of pump runtime.

https://doi.org/10.1051/e3sconf/20185702002

The power value for the pump and compressor works at the same time, it starts at 17:29:44 (figure 7) and ends at 17:29:44 (figure 8), thus determining a time period of three minutes, with an average power value of 9.015[kW].

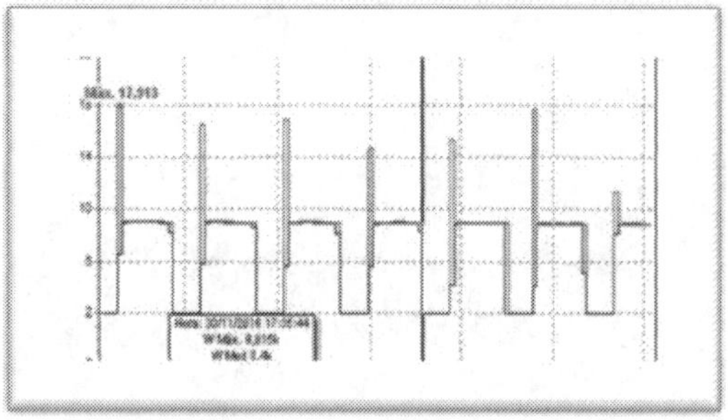

Fig. 8. End of pump and compressor operating time.

Analyzing figure 8, the duration of each cycle is 9 minutes, of which 66.66% of the time (0.6666 h) is spent on the compressor and pump, while the remaining 33.33% only works the pump (0.3333 h). With this data, the estimated energy consumption corresponding to the compressor and pump of the KOOLANT KOOLERS chiller is calculated.

$$E_{CKOL} = Tiempo * Potencia \qquad (5)$$

$$E_{CKOL} = 5,417\ [kWh]$$

In the same way, we proceed to calculate the power demanded by the chiller KOOLANT KOOLERS pump.

$$E_{BKOL} = Tiempo * Potencia \qquad (6)$$

$$E_{BKOL} = 0,647\ [kWh]$$

The total energy consumed in one hour will then be:

$$E_{KOLcal} = E_{CKOL} + E_{BKOL} \qquad (7)$$

$$E_{KOLcal} = 6,065\ [kWh]$$

To verify the data, a comparison is made with the values obtained in the energy analyzer at the same time interval (Figure 9).

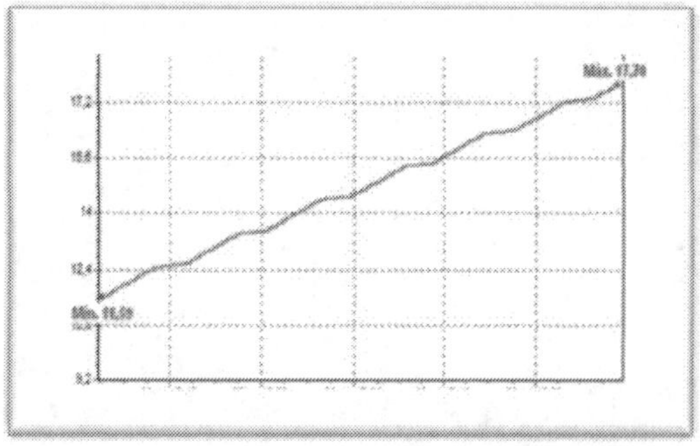

Fig. 9. Energy consumption in one hour of the KOOLANT chiller.

$$E_{KOLdat} = E_{18H00} - E_{17H00}\ [kWh] \qquad (8)$$

$$E_{KOLdat} = 17,703 - 11,59\ [kWh]$$

$$E_{KOLdat} = 6,113\ [kWh]$$

Comparing the value of the calculated energy (E_{KOLcal}) with the energy value obtained by the Power Logger (E_{KOLdat}) confirms that the calculated values are correct. This energy value is taken as the basis for the calculation of the energy consumed monthly by the KOOLANT KOOLERS chiller.

$$E_{KOLmes} = E_{KOLcal} * 24[h] * 30[días]$$

$$E_{KOLmes} = 4366,08\ [kWh/mes]$$

2.3 Analysis of cooling system costs

The monthly energy consumption corresponding to the load of the current cooling system of the production matrices, and the costs associated with this process, depend on the equipment used to accomplish this task. The monthly energy consumption is then projected.

$$E_{SEmes} = E_{HENNmes} + E_{KOLmes} + E_{03mes} \qquad (9)$$

$$E_{SEmes} = 14929,34\ [kWh/mes]$$

2.4 Analysis of energy consumption already implemented the new cooling system

The energy readings corresponding to the cooling equipment were carried out in three connection configurations. These readings were performed with the AEMC 3945 power analyzer. (Figure 10)

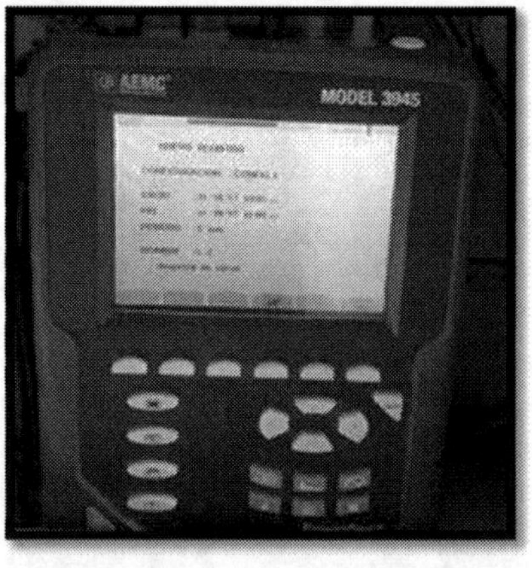

Fig. 10. Energy Analyzer AEMC 3945.

The reading corresponding to the energy consumption of chiller 01 (TAE EVO 201), started at 18:20 on 15/11/2017 and ended at 08:30 on 16/11/2017. (Figure 11)

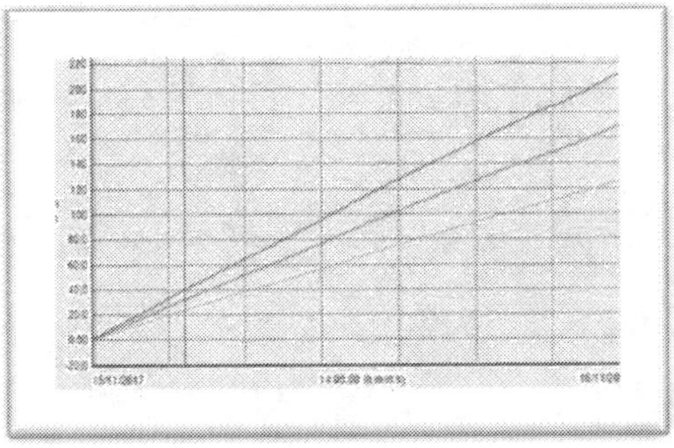

Fig. 11. Energy consumption curve chiller TAE EVO 201.

As a result, an average consumption value of 12.45[kWh] was obtained, and by performing an energy consumption analysis with the same considerations as the analysis of energy consumption prior to redesign, it is obtained:

$$E_{01}=12,45\ [kWh]$$
$$E_{01mes}=12,45*24[h]*30[días]$$
$$E_{01mes}=8964[kWh/mes]$$

The reading corresponding to the energy consumption of chiller 02 (TAE EVO 161), started at 09:05 on 16/11/2017 and ended at 12:55 on 16/11/2017. (Figure 12)

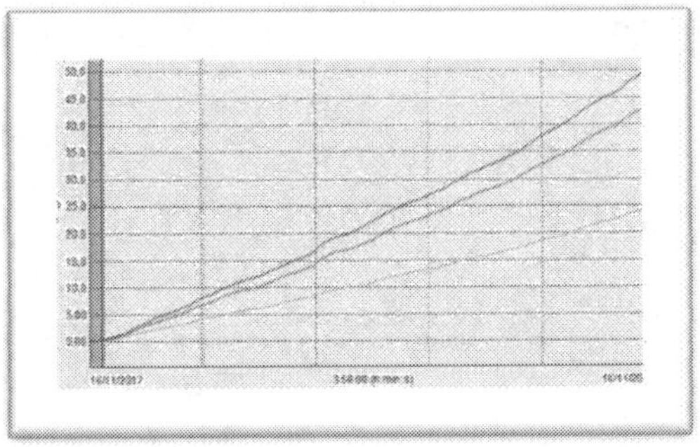

Fig. 12. Energy consumption curve chiller TAE EVO 161.

As a result, an average consumption value of 10.89[kWh] was obtained, and by performing an energy consumption analysis with the same considerations as the analysis of energy consumption prior to redesign, it is obtained:

$$E_{02mes}=10,89*24[h]*30[dias]$$
$$E_{02mes}=7846.2\ [kWh/mes]$$

Chiller 01 and chiller 02 configuration on: the reading corresponding to the energy consumption of chillers 01 and 02, started at 13:05 on 16/11/2017 and ended at 15:05 on 16/11/2017. (Figure 13)

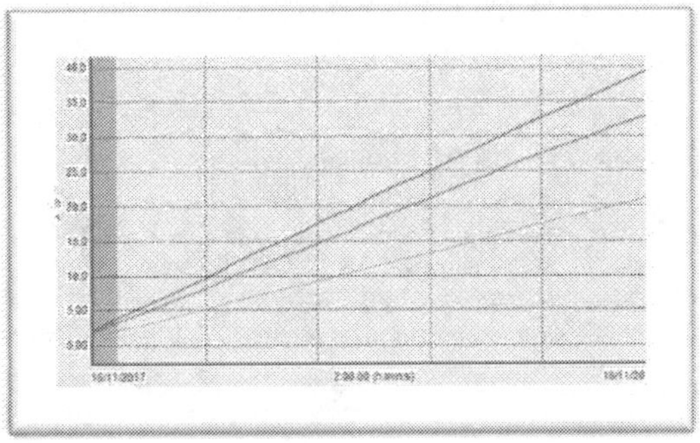

Fig. 13. Energy consumption curve chiller 01 and chiller 02.

As a result, an average consumption value of 16,775[kWh] was obtained, and by carrying out an energy consumption analysis with the same considerations as the analysis of energy consumption prior to the redesign, it is obtained:

$$E_{01y02}=16,775[kWh]$$
$$E_{01y02mes}=16,775*24[h]*30[días]$$
$$E_{01y02mes}=12078[kWh/mes]$$

2.5 Validation of the hypothesis

To validate the hypothesis, it is necessary to compare the results obtained in the previous analysis of energy consumption and the results obtained in the analysis after implementing the new cooling system, for this purpose the following operation is carried out:

$$Ahorro=(1-(12078\ [kWh/mes])/(14929,34[kWh/mes]))\times 100\%$$
$$Ahorro=19,09\ \%$$

As we can see, there is an average saving of 19% which means that the reduction in energy consumption is significant.

3 Contribution of the study to the field

The implementation and study of the automation of a chillers cooling system by means of simulations with the computational fluid dynamics tool, makes it possible to study cases for new refrigerant liquids, as well as to determine better finishes for systems or products that use chillers as part of the cooling system. Industrial automation undoubtedly helps industrial processes to be optimized both in production costs and energy efficiency, with the automation of the cooling system achieved a considerable reduction in energy and thus contributing to the saving of energy resources.

This publication contributes to the further implementation of industrial process automation as it optimizes both environmental and energy resources

4 Conclusions

With the implementation of the new automatic cooling system for the dies of plastic blow moulding machines, a saving of 19% in the energy consumption of the chillers was achieved.

From the location of the chillers in a centralized machine room, it allowed to reduce the ambient temperature in the production area to an average of 20°C, and thus the workers develop their activities in a cooler and more suitable climate.

By recording the time of the plastic blow moulding machines, the optimization of the production times was achieved, thus achieving a reduction in the production cost of plastic containers

References

1. Becka y P. Gohneraa, *Models for Model-BAsed User-Centric Energy Analysis of Industrial Automation Systems*, *ELSEVIER,* vol. 6, **pp. 423-428,** (2011).
2. OECD/IEA, *World Energy Outllok 2009. Executive Summary*, 15 Mai (2011). [En línea]. Available: http://www.iea.org/Textbase/npsum/weo2009sum.pdf.
3. L. A. Barroso y U. Holzle, *The Case for Energy-Proportional Computing, IEE Transactions on Computers,* vol. 40, nº 12, **pp. 33-37,** December (2007).
4. P. Gohner, *What is Industrial Automation?*, 15 MAi (2011). [En línea]. Available: http://www.ias.uni-stuttgart.de/ia/unterlagen/umdruck/chapter01.pdf.
5. H. Vieritz, F. Yazdi, N. Jazdi, D. Schilberg, S. Jeschke y P. Goehner, *Discussions on accessibility in industrial automation systems, IEEE International Symposium on Applied Machine Intelligence and Informatics,* (2011).
6. X. Gong, T. De Pessemier, W. Joseph y L. Martens, *A power data driven energy-cost-aware production scheduling method for sustainable manufacturing, IEEE International Conference on Emerging Technologies and Factory Automation, ETFA,* vol. 2016, nº 7733716, (2016).
7. S. WR, W. RD, B. CE y B. JV, *Conforman Cooling Versus Conventional Cooling: An Injection Molding CAse Study With P-20 anda 3DPTM-Processed Tooling, MRS Proc,* vol. 625, **p. 51,** (2000).
8. S. Z, S. S, Z. AM, G. MF y S. RM, *Potential of Conformal Cooling Channels in Rapid Heat Cycle Molding: A Review, Adv Plym Technol,* **p. 33,** (2014).
9. A. KM y Y. Km, *Conformal cooling channel design and CAE simulation for rapid blow mould, Int J Adv Manuf Technol,* vol. 66, **pp. 11-24,** (2013).
10. R. Holker, A. Jager, B. Khalifa y A. Tekkaya, *Controlling heat balance in hot aluminium extrusion by additive manufactured extrusion dies with conformal cooling channels, Int J Precis Eng Manuf,* vol. 14, nº 1487, **p. 93**, (2013).
11. A. Armillotta, R. Baraggi y F. S, *SLM tooling for die casting with conformal cooling channels, Int Adv Manuf Technol,* vol. 71, nº 573, **p. 83**, (2014).
12. W. Bader y J. Kissock, *Exergy Analysis of Industrial Air Compression, National Industrail Energy Technology Conference,* vol. 22, (2000).

Design and Implementation of a Parabolic Cylinder Collector with Solar Tracking to obtain hot water

Wilson E. Sánchez[1], Mario P. Jiménez [1], Carlos A. Mantilla[1], José M. Toro[1], Miguel A. Villa[2], Germánico Sinchiguano[3]

Departamento de Eléctrica y Electrónica [1], Departamento de Ciencias Exactas [2], Unidad de Gestión de Tecnologías [3], Universidad de las Fuerzas Armadas ESPE, ID: 60104598, Sangolquí, Ecuador

Abstract. This investigation describes the design and implementation of a parabolic trough solar collector (PCC) with solar tracking to obtain hot water. The solar radiation available at the installation site is analyzed, followed by the design and construction of the mechanical system, making a series of calculations for the dimensioning of the reflective base, and a stress and deflection analysis of the structure is performed to verify the feasibility of the design in the ANSYS software. An analysis of the solar tracking system is performed, which is dimensioned from the PCC structure to determine the type of solar tracker to implement; The charging system, consisting of a solar panel and a battery, is dimensioned for the power supply of the tracking system; as a last point, for the heating system is determined the amount of water that is able to heat the system from the energy analysis at the installation site, the heating system is based on placing a Heat Pipe, in the focus of the parabola to receive the solar rays reflected by the collector and heat exchange to the water from a thermowell where the heat pipe condenser enters, finally tests are carried out in the PCC implemented obtaining a global efficiency of 16.37%.

Keywords. Solar collector, PCC, solar tracking, solar panel, hot water

1 Introduction

The use of parabolic trough collectors can be found in different countries that promote the use of renewable energies, such as Mexico, Argentina or Paraguay as well as in Europe. one of the cases is the Mexican Eduardo Zarza's project, with his project theme "Direct generation of steam with parabolic trough collectors" he notes that "solar radiation is never high when the angle of rotation of the collectors are small" [1], he also concluded that "in no case of temperature differences have obtained greater than 41K" [1]. In other works it is proposed that the parabolic trough collector becomes more efficient when installing the solar tracking device, because, first of all, the movement of the collector is continuous, thus describing a trajectory very close to that of the sun, which guarantees a better use of solar radiation [2], is related to the project outlined in this document, which aims to design and implement a parabolic trough collector with solar tracking to obtain water. On the other hand, currently in Ecuador, parabolic trough collectors have not been a subject of study and technological innovation.

1.1 Solar radiation

Radiation is electromagnetic waves which produce the transfer of energy, this occurs in all directions directly from the source [3].

1.2 Parabolic trough solar collector

The main characteristic of PCC's is that they have a linear focus and transform the direct solar radiation into thermal energy when a fluid is heated through a tube located in the focal axis of the same [4], its main components are:

- Parabolic cylindrical reflector: It is characterized by being a curved mirror, in the form of a parabola, which reflects and concentrates on the receiver tube the direct solar radiation that affects its area [5].
- Absorption tube or receiver: This is the core of the PCC, since the performance of the PCC itself depends on the thermodynamic capacity of the absorber [5].
- Solar tracking system: usually from east to west, i. e. the reflective base follows the sun during the day [6].

2 PCCs DESIGN

2.1 Reflective base

https://doi.org/10.1051/e3sconf/20185702003

For the design of the reflective base, bright aluminium with a reflection coefficient of 0.853 was used to reflect the rays of the sun, whose dimensions are 1.75x1.20 meters. Calculations will be made for two focal lengths [7] of 25 and 35 centimeters. The length of the curve is determined by equation (1):

$$S = \int_a^b \sqrt{1 + f'(x)}dx \qquad (1)$$

Already knowing the length of the curve $S = 0.60\ m$; we find the derivative $f'(x)$ of the equation of the parabola, which is:

$$f(x) = y = \frac{x^2}{4 \times F} \qquad (2)$$

Knowing the value of Xmax we can know the value of Ymax, replacing Xmax in the equation of the curve [8], obtaining the values that are represented in the table 1:

Table 1. Relationship Of Focal Lengths

Foco	*Xmáx*	*Ymáx*
0,25	0,49	0,24
0,35	0,51	0,19

The aperture angle α and distance ρ corresponding to the focal point is calculated with the parabola [9]. To calculate α proceed to the following calculation:

$$tg\left(\frac{\alpha}{2}\right) = \frac{Xmax}{F - Ymax} = \frac{0,51}{0,35 - 0,19} = 3,1875$$
$$\alpha = 2 \times arctg(3,1875) = 145,164°$$
$$\rho = \sqrt{Xmáx^2 + (F - Ymáx)^2}$$
$$\rho = \sqrt{0,51^2 + (0,35 - 0,19)^2} = 0,53\ cm$$

2.2 Receiver tube

Now we proceed to calculate the receiver tube diameter using the following equation:

$$\emptyset min = \frac{2 \times \rho \times tg16'}{cos\left(\frac{\alpha}{2}\right)}$$
$$\emptyset min = \frac{2 \times 0,53 \times tg16'}{cos\left(\frac{145,164}{2}\right)} = 1,6\ cm$$

Using the value of focal length equal to 25 cm, a minimum pipe diameter value of 22 centimetres is obtained [10]; therefore the focal length to be used in the project will be 35 centimetres.

2.3 Water heating

It is necessary to calculate the heat that will be received by the absorber tube coming from the reflective surface of the PCC, for this purpose a prototype of the system to be implemented to obtain the temperature values, as well as the necessary data for tank sizing [11]. With which the coefficient of losses in the tank is calculated, obtaining:

$$Uc = 5,497\ W/m^2K$$

The area of the receiver tube is obtained by the following calculation [12]:

$$A_r = \pi \times 0,047 \times 1,80 = 0,265\ m^2$$

So the heat incident on the heat pipe is:

$$Q_{inc} = 486,19\ W$$

This data is used to calculate the minimum water mass that can be heated up to 60 °C. As the heat pipe is located inside a thermowell for heat exchange in the tank, losses in the tank [13] are taken into account:

$$Q_{perdidas} = 62,5\ W$$

This means that the useful heat delivered by the collector to the water stored in the tank must be

$$Q_{util} = Q_{inc} - Q_{perdidas}$$
$$Q_{util} = 486,19 - 62,5 = 423,69\ [W]$$

Where the necessary heat

$$Q_{nec} = Q_{inc} \times 3600\ s = 423,69\ \times 3600$$
$$Q_{nec} = 1525,28\ kJ$$
$$Q_{nec} = m \times Cp \times (Tf - Ti)$$
$$1525,28 = m \times 4,18 \times (60 - 15,65)$$
$$m = 8,22\ kg$$

Therefore, in one hour the PCC is capable of raising the water temperature from 15.65°C to 60°C from 8.22 liters of water, so the tank to be designed will be cylindrical and will have a capacity of 10 liters.

A water recirculation system is not necessary for the operation of the heating system because this water heating system was designed using the heat pipe heat exchange system, where the heat pipe condenser enters the storage tank through a thermocouple, which is where the heat exchange with the water takes place.

3 PCC Implementation

As soon as the design conditions for the calculated and simulated stress and deflection parameters in the ANSYS software are met. The system was built and implemented in the laboratory of Renewable Energies of the Universidad de las Fuerzas Armadas ESPE, Fig. 1, Fig. 2.

Figure 1. PCC constructed

Figure 2. Built in heating system

The solar tracking is on one axis due to being near the equatorial line the solar declination in the solstices does not exceed 23.4° according to the astronomy simulation software of the University of Nebraska-Lincoln with which the solar movement on Latacunga was analyzed, therefore a north-south tracking is not necessary.

3.1 Power supply unit

The necessary equipment was sized to power the tracking system based on its technical characteristics in order to have autonomy in the system.

3.2 Battery

The minimum capacity of the battery to be implemented in the project is calculated based on the technical characteristics of the solar tracker.

$$Capacity = \frac{E_B}{V} = \frac{135\ [Wh]}{12\ [V]}$$
$$Capacity = 11{,}25\ Ah$$

Therefore, a 12 V battery with a capacity of at least 11.25 Ah is required. A solar charge controller of 5[A] is implemented in the system. In order to calculate the solar panel, the power required is 135Wh, so the energy of the solar field is:

$$E_{CS} = \frac{E_d}{\eta_B}$$

$$E_{CS} = \frac{108}{0{,}8} = 135\ [Wh]$$

With this you get the power of the panel will be:

$$E_p = \frac{E_{CS}}{Sun\ peak\ hours\ (HSP)}$$
$$E_p = \frac{135}{5} = 27\ [W]$$

With this value, the solar panel must have a power greater than or equal to 27[W], so for this project is implemented a panel of 30[W].

4 Results of the tests

For the tests, initially the temperature of the water entering the tank was measured, the temperature at which the environment was found. During the tests, an eighth of a covered sky was obtained.

The measurements obtained are shown in Fig. 3, where the solar radiation on the installation site is represented throughout the test, and in Fig. 4, the water temperature in the tank versus time.

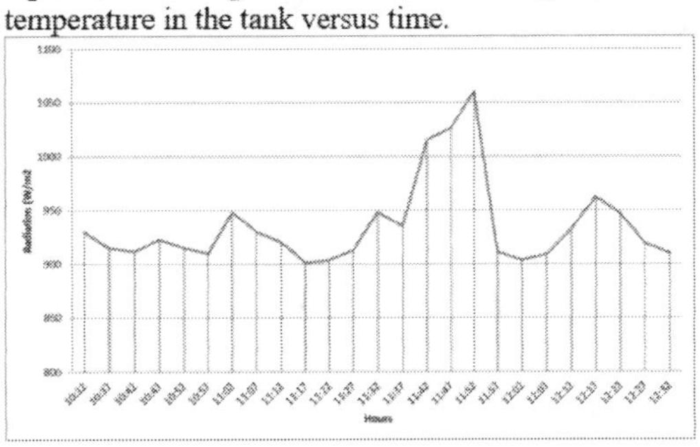

Figure 3. Solar radiation vs. time June 6

In Fig. 4, it is possible to observe solar radiation versus time, the incident energy of the sun during the two hours is 1873.25 Wh/m2.

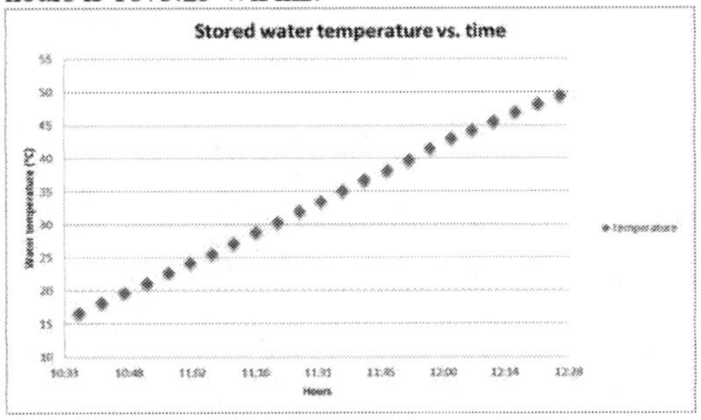

Figure 4. Storage water temperature vs. time June 6

Fig. 4, shows that the temperature of the water in the tank (represented on the y-axis) increases from 15°C to 50.2°C in two hours (the time is represented on the x-axis of the graph).

This means that 1471.35[KJ] of useful energy was produced during two hours through the implementation of the Heat Pipe heating system in the reflector base of the PCC, which allowed an increase in the temperature of the water stored in the tank of 35.2°C. With the data obtained in practice, useful energy is calculated [14] based on the data obtained in practice, obtaining:

$$Q_{util} = m \times Cp \times (T_f - T_i)$$
$$Q_{util} = 10 \times 4{,}18 \times (50{,}2 - 15)$$
$$Q_{util} = 1471{,}36\ [KJ]$$
$$Q_{util} = 1471{,}36\ [KJ] \times \frac{1}{3600\ s} = 408{,}71\ [Wh]$$

The solar irradiation in the two hours of operation of the system is 1873.25 (Wh/m2), so the energy captured by the collector will be of:

$$Q_{in} = Sc \times E_{sol} \times \cos\theta$$

Sc being the useful capturing surface of the reflector base with solar tracking, and θ the angle of solar incidence at the place of installation of the collector.

$$Q_{in} = 1{,}5418 \times 1873{,}25 \times 0{,}8643$$
$$Q_{in} = 2496{,}25\ [Wh]$$

With this we will have a global efficiency [15] of:

$$\eta = \frac{Q_{util}}{Q_{in}}$$
$$\eta = \frac{408{,}71}{2527{,}23}$$
$$\eta = 0{,}1637 = 16{,}37\%$$

Therefore the system has a global efficiency of 16.37%.

5 Conclusions

A parabolic trough collector was designed, oriented to the north with an inclination of 6° with respect to the horizontal one with east-west solar tracking controlled by a photovoltaic sensor; by means of the implementation of the Heat Pipe heating system in the focus of the reflective base of the PCC 1471,35 [KJ] of useful energy was produced during two hours, which allowed to have an increase of temperature of the water stored in the tank of 35,2°C.

The overall performance of the system was 16.37% since the solar incident energy on the collector was 2527.23 Wh and the useful energy used by water to raise its temperature from 15°C to 50.2°C was 408.71[Wh].

Glossy aluminium with a coefficient of reflection of 0.853 was used for the reflective base in order to reflect the sun's rays and, at the same time, low weight; the structure of the system was constructed of A36 steel and painted in the oven to withstand adverse climatic conditions, the storage tank is constructed of stainless steel.

The solar tracking system has autonomy for its operation through its power supply, which was designed by means of a solar panel and a battery, because it has a DC motor that moves an actuator by means of gears at a speed of 5.7 mm/s and has a protection class IP54 for the actuator and IP65 for the control box.

By means of a four-channel digital thermometer, to which type K thermocouples are connected, the water temperature is measured at the tank inlet, in the water stored inside the tank and in the Heat Pipe condenser; solar radiation is measured by means of a pyranometer during system operation.

References

1. E. Zarza, "Generación Directa de Vapor con Colectores Solares Cilindro Parabólicos Proyecto DIrect Solar Steam (DISS)," Sevilla, 2003.
2. J. Fabara, Desarrollo experimental de un sistema de seguimiento solar para la obtención de la máxima radiación en concentradores cilíndrico parabólicos, Sangolquí: ESPE, 2016.
3. P. D. Tagle, A. Agraz y C. Rivera, «Study of applications of parabolic trough solar collector technology in Mexican industry,» Instituro Tecnológico y de Estudios Superiores, Monterrey, 2016.
4. A. Z. Fafez, A. M. Attia, H. S. Eltwab, A. O. ElJousy, A. A. Afifi, A. G. AbdEihamid, A. N. AbdElqader, S. Fateen, K. A. ElMetwally, A. Soliman y M. Ismail, «Design analysis of solar parabolic trough thermal collectors,» India, 2017.
5. C. Echeverria, «Diseño de un colector cilindro parabólico compuesto con aplicación para el calentamiento de agua,» Universidad de Pirhua, Piura, 2011.
6. J. Macedo Valencia y J. Ramirez Avila, «Design, construction and evaluation of Parabolic trough collector as demostrative prototype,» Quintana Roo: Departamento de Ingeniería, 2013.
7. C. Paredes, «Diseño de captador solar cilíndrico parabólico para aplicaciones rurales en Paraguay,» Asunción, 2012.
8. F. Mahjouri, «Vacuum Tube Liquid-Vapor (Heat-Pipe) Collectors,» Thermo Technologies, 2015.
9. Y. Garcia Ortiz, J. Yañez Mendiola y L. Valenzuela, «Cylindrical parabolic collectors material from low cost (stainless steel) applied to dried hybrid energy system,» Dyna, 2016.
10. J. A. Carta Gonzales y R. Calero Pérez, Centrales de energías renovables, Madrid: Pearson, 2013.
11. B. Nafiz Ahmed y E. Farzana Mollick, «Development and performance test of an evacuated tube solar water heater,» Department of Mechanical Engineering Khulna University of Engineering & Technology, Khulna-9203, Bangladesh, 2016.
12. R. Cuestas y I. Rincón, «Diseño y construcción y evaluación operacional a pequeña escala de un potabilizador solar parabólico para agua,» Bucaramanga, 2011.

https://doi.org/10.1051/e3sconf/20185702003

13. F. Incropera, Fundamentos de transferencia de calor, Mexico D.F.: Prentice Hall Hispanoamericana, 1999.
14. R. Cherian y L. G. Lasithan, «Performance enhancement of solar water heater using compound parabolic reflector and numerical simulation of thermal lossees,» College of Engineering, India, 2016.
15. Y. Cengel, Transferencia de calor y masa, Ciudad de México: McGraw Hill, 2007.

https://doi.org/10.1051/e3sconf/20185702004

Economic Analysis of Waste Heat Utilization for Coal-fired Power Units

Wang Kai , *Lu* Kun, *Wang* Shuai

Shandong Branch, Huadian Electric Power Research Institute Co., Ltd, Jinan, China

Abstract. Strict energy conservation policies require power plants to minimize energy consumption as much as possible. Low pressure economizer is an effective method to recycle waste heat from boiler exhaust gas. In this paper, the economic analysis model of coal-fired power plant is established by means of equivalent enthalpy drop method, and the waste heat utilization effect on thermal systems is investigated, including boiler efficiency, steam turbine power output, condensate pump, induced draft fan, electro static precipitator and desulfurization system. Then the economic effect on a 1000 MW ultra super critical unit in Shandong Province is calculated. The results show that waste heat utilization could comprehensively reduce standard coal consumption rate by 1.89 g/(kW·h), and the unit could save about 10.8 thousand tons standard coal and 1.29 billion dollars every year. Besides, the energy saving effect is more significant for lower load condition.

1 Introduction

At present, more intense emission requirements are raised for coal fired power plants because of the improvement of energy conservation and emission reduction policies. According to statistics [1,2], the boiler exhaust loss accounts for the largest proportion of boiler heat loss, and the designing smoke exhaust temperature ranges from 120 ℃ to 140℃ for most power plants.

However, the actual smoke exhaust temperature is generally higher than the design data. Take the flue pipe corrosion out of consideration, if the boiler exhaust temperature could be reduced to 70℃~90℃, then this part of waste heat could be used in steam turbine regenerative system [3].

Therefore, the power generation efficiency will be increased by 2 to 4 percent, and the standard coal consumption will reduce by 2 to 4 g/(kW·h) [4~6]. The thermal power generation capacity is about 4.29 trillion kW·h in 2016 in China, if exhaust waste heat is applied to all the national thermal power plants, the annual coal consumptions could be saved about 8.6 to 17 million tons [7].

The thermal economic evaluation for low pressure economizer should consider not only the waste heat recovery amount, but also the work capacity of squeezing out steam after being returned to the turbine [8]. Two methods are currently being used to evaluate the energy saving effect for low pressure economizer.

One is adopting the heat balance diagram, namely executing two tests for steam turbine thermal performance with and without the economizer runtime respectively [9], in order to get the two tests difference of power supply coal consumption rate.

The other one is using the equivalent enthalpy drop method to calculate the energy saving effect by means of steam turbine thermal performance test [10].

As to the research of waste energy recovery utilization, many scholars analyzed its energy saving effect by means of equivalent enthalpy drop method and thermal balance method. Huang analyzed the waste heat utilization effect of flue gas by equivalent enthalpy drop method for various utilization schematic systems [11].

Some studies show that exhaust waste heat utilization can increase the power output of the steam turbine, and the energy saving effect could be optimized by adjusting the low pressure economizer arrangement and the heat transfer parameters [12~15].

However, most studies at present focus mainly on the optimal design of the low pressure economizer, and on the local benefits or influence of waste heat recovery, few has investigated the overall effect of waste heat utilization on the unit system.

Therefore, the comprehensive heat recovery effect on the entire system of a 1000MW ultra super critical coal-fired unit was analyzed in this paper, including boiler efficiency, steam turbine power output, draft fan, condensation pumps, dust collector and desulfurization system. Besides, the load effect on energy saving performance and thermal economy is discussed.

2 Calculation model

A 1000MW ultra super critical coal-fired unit in Shandong Province was investigated, which include an ultra-super critical once-through boiler with the design coal is composed of : Ash-8.8%, Carbon-56.3%, Hydrgen-3.8%, Sulphur-0.2%, Nitrogen-0.8%, Oxgen-

 https://doi.org/10.1051/e3sconf/20185702004

12.1%, Moisture-18.0%. The low calorific value of design coal is 21MJ/kg.

In addition, the steam turbine belongs to the condensing type with the model type N1000-26.3/600/600, and eight stage heat extraction is adopted to heat condensation water.

The low temperature electrostatic dust collection technology is used on this unit, and the low-pressure economizer is used to recycle the exhaust waste energy and heat condensation water.

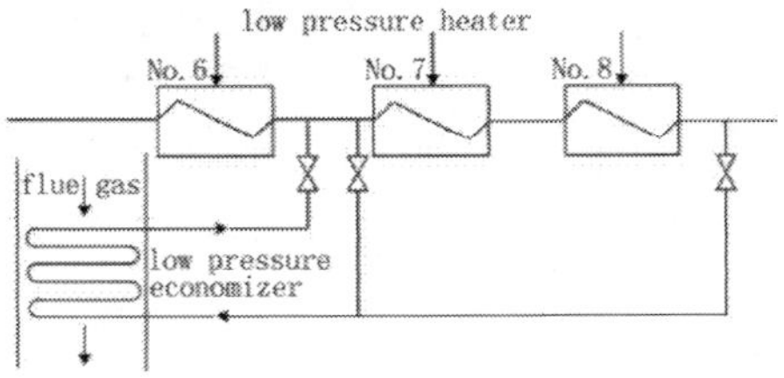

Fig. 1. Schematic diagram of low pressure economizer arrangement.

The schematic diagram is shown in Fig. 1, which is as follows. Low-pressure economizer inlet water is from No.7 low pressure heater outlet and shaft seal water valve export. After that, it enters into economizer module, and exchange heat with flue gas. Then it flows out of the module and returns into the No.6 low pressure heater inlet.

3 Waste heat utilization impacts on thermal systems

3.1 Effect on boiler efficiency

Boiler thermal efficiency is determined by heat balance and thermodynamic calculation of low pressure economizer. The detailed calculation process is shown in Fig. 2, and the heat exchange Q_j is calculated as the initial heat transfer in the next calculation.

Besides, the exhaust flue gas temperature t_j, outlet water temperature t_{wj}, average heat transfer temperature difference Δt_j and economizer heat transfer Q_j are calculated in turns.

If the criterion $|Q_j-Q_{j-1}|/Q_j<\varepsilon$ (calculation error) is not met, then perform the next iterative calculation. By modifying the method of iteration, the efficiency of calculation can be improved.

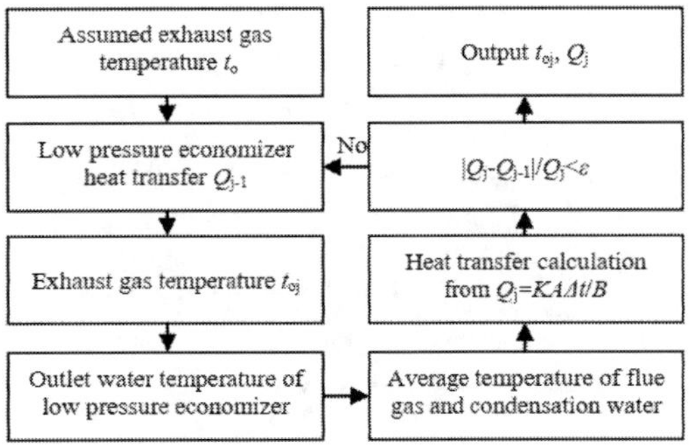

Fig. 2. Calculation process for heat exchange.

In turbine heat acceptance conditions, the calculation parameters are as follows: exhaust gas inlet temperature-130℃, condensation water inlet temperature-70.2℃, gas flow aera-338 m^2, heat transfer aera-77500 m^2.

After iteration and verifying calculation, the thermal-dynamic performance parameters of low pressure economizer are shown in Table1.

Table 1. Thermal Performance of Low Pressure Economizer.

Items	Data
Exhaust gas outlet temperature / ℃	86.5
Condensation water outlet temperature / ℃	109.8
Heat transfer coefficient / W/(m^2·K)	36.2
Average heat transfer temperature difference / ℃	18.2
Heat exchange amount	51153Kw

It can be seen from Table 1 that the low pressure economizer could decrease boiler exhaust temperature from 130 ℃ to 86.5 ℃, and the exhaust heat loss decreases from 6.9% to 4.5%. Therefore, boiler thermal efficiency rises from 93.5% to 95.8%.

3.2 Effect on steam turbine power output

The equivalent enthalpy drop method was used to analyze the influence of the waste heat utilization on steam turbine. During the heat recovery process, the condensation water could be heated, which can reduce the turbine steam extraction amount and generate extra power.

Based on the equivalent enthalpy drop method, the equivalent enthalpy drop increment ΔH caused by the exhaust heat utilization could be calculated, and the heat recovery performance could be evaluated.

If the low pressure economizer is in series arrangement as shown in Fig. 3, the equivalent enthalpy drop increment ΔH is as follows in (1).

$$\Delta H = Q_d \cdot \eta_j / D_0 \quad (1)$$

Where D_0 is superheated steam flow rate, kg. Q_d is waste heat amount, kJ. H_j is extraction efficiency of the reheating heater corresponding to the waste heat, %.

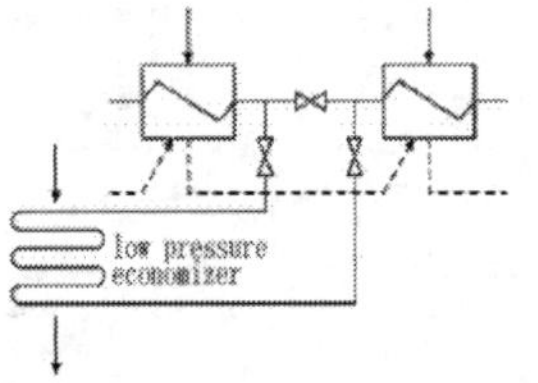

(1008) Parallel arrangement

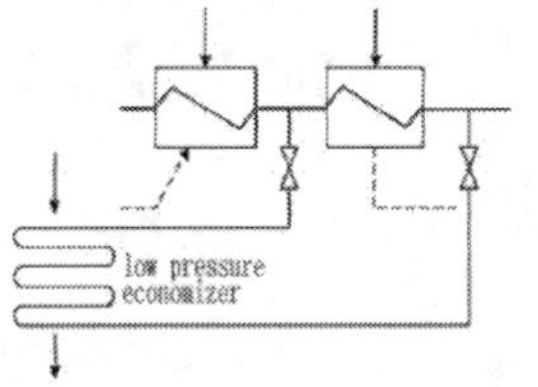

(b)Series arrangement

Fig. 3. Typical arrangement patterns for low pressure economizer.

If the low pressure economizer is in parallel arrangement with the No. n to (m-1) low pressure heater, the equivalent enthalpy drop increment ΔH, heat rate decrement Δq and standard coal consumption rate decrement Δb_{cp} could be calculated as follows.

$$\Delta H = \beta[(h''_d - h_{w(m-1)})\cdot \eta_m + \sum^{m-1}_{j=n}(\tau_j \cdot \eta_j)] \quad (2)$$

$$\Delta q = q \cdot \Delta H / (H + \Delta H) \quad (3)$$

$$\Delta b_{cp} = 1000 \cdot \Delta q / (\eta_b \cdot \eta_p \cdot 29270) \quad (4)$$

Where β is discharge coefficient of inlet condensation water for low pressure economizer. H_d" is outlet water enthalpy, kJ/kg. $h_{w(m-1)}$ is outlet water enthalpy of No. (m-1) low pressure heater, kJ/kg. η_m is extraction efficiency of No. m heater, %. I_i is enthalpy increase of parallel heater condensation water. Q is the original heat rate of steam turbine, kJ/(kW·h). H is the net equivalent enthalpy decrease, kJ/(kW·h). η_p and η_b are pipe efficiency and boiler efficiency respectively.

Because the energy saving effect of the waste heat is achieved by squeezing out steam from the steam turbine, the boiler efficiency is defined as the air preheater outlet gas temperature.

Table 2 shows the extraction parameters of a 1000 MW steam turbine in heat acceptance condition. By means of equivalent enthalpy method, the waste heat utilization effect on steam turbine is shown in Table 3.

It can be seen that power generation increases 11.46 kJ per kilogram steam, and the turbine power output increases 8.4 MW with 2.32 g/(kW·h) decrease for standard coal consumption rate.

Table 2. Thermal Parameters of All Heaters in Turbine Heat Acceptance Conditions.

Heater Number	Steam Extraction temperature /℃	Steam Extraction Enthalpy /kJ·kg^{-1}	Extraction Flow Rate /kg·s^{-1}	Outlet Water Temperature /℃
1	395.4	3136.4	39.98	289.9
2	353.2	3064.2	75.79	268.7
3	481.9	3420.9	31.28	219.8
Deaerator	382.5	3223.7	70.66	189.6
5	289.4	3040.5	30.74	153.1
6	191.9	2851.6	35.40	121.8
7	99.4	2629.4	21.17	83.0
8	94.3	2495.3	20.95	60.9

Table 3. Waste Heat Utilization Effect on Steam Turbine.

Items	Before Modification	After Modification
Exhaust gas temperature / ℃	130.0	86.5
Squeezing steam work of No.8 heater ΔH_8 / kJ·kg^{-1}	/	0.82
Squeezing steam work of No.7 heater ΔH_7 / kJ·kg^{-1}	/	1.48
Squeezing steam work of No.6 heater ΔH_6 / kJ·kg^{-1}	/	9.15
Increment of equivalent enthalpy drop ΔH / kJ·kg^{-1}	/	11.43
Steam equivalent enthalpy drop H / kJ·kg^{-1}	1357.62	1369.08
Heat consumption rate q_0 / kJ·(kW·h)$^{-1}$	7346	7283
Decrement of heat consumption rate Δq_0 / kJ·(kW·h)$^{-1}$	/	63
Steam consumption rate d_0 / kg·(kW·h)$^{-1}$	2.708	2.689
Generation standard coal consumption rate b_{cp} / g·(kW·h)$^{-1}$	277.51	275.19
Standard coal consumption savings Δb_{cp} / g·(kW·h)$^{-1}$	/	2.32
Power generation P_e / MW	1000	1008.4

3.3 Effect on condensate pump

If the low pressure economizer is arranged in series, a resistance is added to the original system, and it is necessary to increase condensate pump output or use booster pump to overcome the resistance.

On the contrary, the power consumption will not increase if the economizer is arranged in parallel, because the condensation water pressure is sufficient to overcome the economizer resistance.

As to the series arrangement, the power consumption increment of condensate pump ΔP_p is equal to $q_w \cdot \Delta P_w/(3600 \cdot \eta_p)$. Where q_w is economizer water flow rate, m^3/h. ΔP_w is economizer water resistance, Pa. η_p is condensate pump efficiency, %.

Because of the low pressure economizer for 1000 MW ultra super critical unit is not in simple series or parallel arrangement, the booster pump method has some energy saving effect, compared with the increasing condensate pump head method.

However, according to engineering experiences, the booster pump method is with more cost and engineering construction.

Therefore, only the increasing condensate pump head method is analyzed in this paper.

According to the economizer arrangement shown in Fig.1, the water intake could be divided into two parts: No.7 heater outlet and the water valve of shaft seal.

 https://doi.org/10.1051/e3sconf/20185702004

The water from shaft seal valve could be deemed as parallel arrangement without increasing condensate pump output.

In addition, water from the No.7 heater outlet is in series model, so the condensate pump output increases to overcome economizer resistance. The power consumption result is shown in Table 4.

Table 4. The Power Consumption Increment of Condensate Pump and Draft Fan.

Items	Data
Pump efficiency η_p / %	84.5
Water flow rate from No.7 heater outlet / $m^3 \cdot h^{-1}$	790
Water flow rate from shaft seal valve / $m^3 \cdot h^{-1}$	336
Main condensation water flow rate / $m^3 \cdot h^{-1}$	1875
Water resistance of low pressure economizer ΔP_w / Pa	612
Power consumption increment ΔP_p / kW	226.5
Induced draft fan efficiency η_f / %	83.8
Flue gas flow rate q_y / $m^3 \cdot h^{-1}$	4135663
Gas resistance of low pressure economizer ΔP_y / Pa	623
Power consumption increment ΔP_f / kW	854.1

3.4 Effect on flue gas system

Low pressure economizer usually arranges between air preheater and electro precipitator, and it would increase the flue resistance. Hence, the draft fan output will increase to overcome flue gas resistance.

The power consumption of draft fan ΔP_f is equal to $q_y \cdot \Delta P_y/(3600 \cdot \eta_f)$. Where q_y is economizer gas flow rate, m^3/h. ΔP_y is economizer gas resistance, Pa. η_f is induced draft fan efficiency, %.

As to the 1000MW ultra super critical unit, the power consumption increment result for draft fan is shown in Table 4.

In addition, the waste heat utilization has great effect on electro static precipitator performance. According to engineering experiences, the electro precipitator removal efficiency is greatly influenced by gas temperature.

The gas temperature reduction would decrease the flue gas viscosity, which could reduce the internal friction of particle movement and improve the drive speed of charged dust.

Besides, the gas temperature reduction would cause gas density and breakdown voltage increase. Hence, the dust collector efficiency is been improved.

According to the 1000 MW unit investigated in this paper, after the waste heat utilization modification, because of the decrement of flue gas temperature and volume flow rate, the dust particle specific resistance reduces after SO_3 absorption, which in turn greatly improves the electro precipitator removal efficiency.

3.5 Effect on Desulfurization Process

Waste heat utilization reduces the inlet gas temperature of desulfurization system, which causes the temperature decrement of desulfurization reaction and outlet gas.

Therefore, heat recycling will effectively reduce the water consumption of desulphurization system, in which the water consumption with 80~90 percentage concentrated in desulfurization tower.

In addition, the desulfurization reaction temperature decrement improves the dissolution and absorption of SO_2 in limestone slurry with inhibiting the reaction rate. At the same time, the SO_3 concentration in outlet flue gas is reduced because of soot particles absorption effect, which can sharply reduce the acid mist generation and PM2.5 emissions.

Besides, according to the 1000MW unit running experiences, the heat utilization modification did not improve desulfurization efficiency significantly.

4 Annual economic analysis

Combined with the heat utilization effect on steam turbine output and additional power consumption for draft fan with condensate pump, the comprehensive decrement for standard coal consumption rate $\Delta b'_{cp}$ could be obtained as follows.

$$\Delta b'_{cp} = \Delta b_{cp} - \Delta P \cdot q / (P_e \cdot 29.27) \tag{5}$$

Where ΔP is auxiliary engine power consumption, kW. P_e is unit power output, kW. It is found that under turbine heat acceptance circumstances, the heat utilization could comprehensively reduce standard coal consumption rate by 1.89 g/(kW·h).

Moreover, the energy saving effect for economizer would change with the load variation, which is shown in Table 5.

It could be seen from Table 5 that along with the load decrement, the boiler efficiency increment rises and auxiliary engine power consumption decreases.

Table 5. Energy Saving Effect in Various Load Conditions.

Items	100%THA	75%THA	50%THA
Boiler efficiency increment $\Delta\eta_b$ / %	0.63	0.70	0.78
Steam turbine power output increment ΔP_e / MW	8.4	5.1	2.9
Auxiliary engine power consumption increment / MW	1.08	0.56	0.34
Comprehensive standard coal consumption rate decrement $\Delta b'_{cp}$ / $g \cdot (kW \cdot h)^{-1}$	1.89	2.11	2.32

Hence, the energy saving effect of heat utilization is more significant for lower load condition. The lower the load, the more the standard coal consumption decreases.

Based on the above investigation, the annual economic evaluation of the 1000MW unit could be calculated. Assuming that the unit runs 5200 hours a year, and the time proportion of the 100%, 75% and 50% load condition are 35%, 40% and 25% respectively.

Therefore, it can infer that the unit could save about 10.8 thousand tons standard coal and 1.29 billion dollars every year by means of waste heat utilization.

5 Conclusions

By means of equivalent enthalpy drop method, the flue gas waste heat utilization of a 1000 MW ultra super critical unit in Shandong Province is investigated, and its effect on boiler efficiency, steam turbine power output, condensate pump, induced draft fan, electro static precipitator and desulfurization system is analyzed.

It is found that under turbine heat acceptance circumstances, the heat utilization modification would decrease boiler exhaust temperature from 130 ℃ to 86.5℃, and boiler efficiency rises from 93.5% to 95.8% with a 2.32 g/(kW·h) decrement for standard coal consumption rate.

In addition, waste heat utilization would cause power consumption increment for draft fan and condensate pump, precipitator efficiency improvement and desulfurization water consumption decrement. Under turbine heat acceptance circumstances, it could comprehensively reduce standard coal consumption rate by 1.89 g/(kW·h).

Finally, the investigation shows that by means of waste heat utilization, the unit could save about 10.8 thousand tons standard coal and 1.29 billion dollars every year, and the energy saving effect is more significant for lower load condition.

References

1. T. Bai, Z.P. Jin, Y.L. Wang, et al, "Study on the waste heat utilization system of 300 MW CFB boiler," Clean Coal Technology, **23** (2017)
2. G.V.P. Varma, T. Srinivas, "Power augmented steam power plant in a cogeneration cement factory," Journal of Energy Engineering, **143** (2017)
3. Y.F. Ma, L.J. Yang, J.F. Lu, et al, "Techno-economic comparison of boiler cold-end flue gas heat recovery processes for efficient hard-coal-fired power generation," International Journal of Energy Research, **41** (2017)
4. L. Pang, "Influence of a low pressure economizer on the thermal cost-effectiveness of its unit at a constant power output," Journal of Engineering for Thermal Energy and Power, **31** (2016)
5. F.X. Yang, H.Z. Tan, J.J. Zhang, et al, "Comparison of different systems for waste heat recovery from flue gas in a 1000 MW power plant," Clean Coal Technology, **23** (2017)
6. G.C. Liu, Y.F. Liao, X.Q. Ma, "Synthetical analysis of the heat recovery effects on coal-fired power units," Boiler Technology, **48** (2017)
7. X.Y. Huang, F.Z. Sun, Y.T. Shi, "Energy saving effect of additional installed low-pressure economizer in thermal dynamic system of thermal power plant," Thermal Power Generation, **03** (2008)
8. S. Park, J. Seo, "Analysis of air side economizers in terms of cooling energy performance in a data center considering exhaust air recirculation," Energies, **11** (2018)
9. Y. Chen, Z.Q. Bao, "The thermal economy analysis of a 600 MW ultra-super critical unit with low pressure economizer by equivalent enthalpy drop method", Energy Research an Utilization, vol.05, pp. 40-43, 2017.
10. J.X. Wang, W.P. Yan, D. He, "Energy conservation analysis for waste heat utilization system applying air preheater and using primary air to heat condensate water," Thermal Power Generation, **46** (2017)
11. T. Lv, Z.J. Zhang, S.J. Zhang, "Optimized choice for the connection mode of the water side of the low pressure economizer of a 600 MW boiler," Journal of Engineering for Thermal Energy and Power, **28** (2013)
12. L. Wu, "Analysis on transformation of low pressure economizer of 100 MW Unit boiler and its economical efficiency," Guangdong Electric Power, **28** (2015)
13. D. He, W.P. Yan, J.Q. Xu, "Comprehensive optimization of inlet water temperature and temperature rise of low pressure economizer in utility boiler," Boiler Technology, **46** (2015)
14. N.S. Ganesh, T. Srinivas, "Thermo dynamic assessment of heat source arrangements in Kalina power station," Journal of Energy Engineering, **139** (2013)
15. B. Zhao, L.J. Zhao, P. Li, et al, "Thermal economic analysis on low-temperature waste heat recovery," Thermal Power Generation, **44** (2015)

E3S Web of Conferences **57**, 02005 (2018)
ICSREE 2018

https://doi.org/10.1051/e3sconf/20185702005

Synthesis of hexamethylene diisocyanate-functionalized graphene oxide for solar cell applications

J A Luceño[1], *A M* Díez-Pascual[1], *R* Peña[2] and *P* García-Díaz[2]

[1]Analytical Chemistry, Physical Chemistry and Chemical Engineering Department
[2]Signal Theory and Communications Department, Alcalá University, 28871, Madrid, Spain

Abstract. Graphene (G), an allotrope of carbon with exceptional optical, electronic, thermal and mechanical properties, and its oxidized form graphene oxide (GO), show huge potential for a broad range of applications. In particular, their high conductivity, transparency, flexibility, and abundance make them suitable for polymer solar cells (PSCs). However, their insolubility in common organic solvents hinders their applications. Consequently, novel functionalization approaches are pursued. The present work is devoted to the preparation of hexamethylene diisocyante-functionalized graphene oxide (HDI-GO). The synthesized nanomaterial shows a highly hydrophobic nature and can be dispersed in organic non-polar solvents, hence is a prospective candidate to be combined with conjugated polymers for solar cell applications.

1 Introduction

Graphene (G), an allotrope of carbon like diamond, graphite and fullerenes, has attracted a lot interest in recent years both for fundamental studies and potential applications [1]. It is a flat, atomically thick 2D sheet composed of sp^2 carbon atoms arranged in a honeycomb structure (Figure 1). It presents superior electronic, thermal and mechanical properties, very large surface area and the highest electrical conductivity known at room temperature [2]. Besides, G is believed to be the strongest material on earth, with an elastic modulus of 1 TPa [3], tensile strength of ~150 GPa and a breaking strength 200 times higher than steel [4]. Its extremely high carrier mobility, broad absorption spectral range, high optical transparencyand abundance make it a suitable candidate for application in optoelectronic devices suchas displays, light-emitting diodes, and solar cells [5,6]. In this regard, a lot of effort is currently focused onto the development of polymer solar cells (PSCs) with G or its derivatives [7]. They have great potential to be used as transparent conductive electrodes (anode, cathode or both of them), as additives at the interfacial layers (cathode buffer or hole transporting layers) and as acceptor additives to donor-acceptor materials in bulk heterojunction (BHJ) organic photovoltaic cells [8].

Several routes have been reported for the preparation of G, including chemical vapor deposition (CVD) of hydrocarbons onto transition metal surfaces, micromechanicalexfoliation of graphite, epitaxial growth on electrically insulating substrates like SiC wafers, electrochemical intercalation, thermal exfoliation or chemical reduction of graphiteoxide [9,10]. However, these methods lead to very small amount of nanomaterial. Further, G is insoluble in common solvents, which hinders its applications. G can be oxidized following the Hummers´ method [11] to yield graphene oxide (GO, Figure 1), a water-soluble nanomaterial since comprises epoxide, hydroxyl and carbonyl groups on the basal planes and carboxylic acids on the edges. Thus, upon sonication in aqueous media, GO easily exfoliatesinto colloidal suspensions of single layers. However, the hydrophilic nature of GO hinders its dispersion in common organic solvents,which limits the preparation of polymer/GO nanocomposites. To solve this drawback, novel functionalization strategies are pursued.

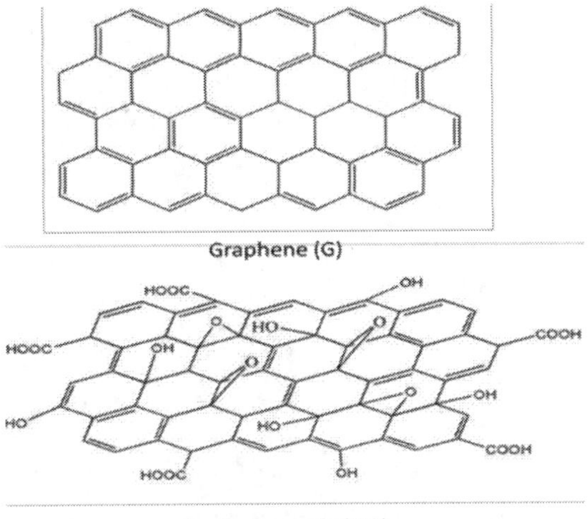

Figure 1. Schematic representation of G and GO

G and GO have been functionalized via covalent and non-covalent approaches [12,13]. When organic molecules are covalently attached onto the G (or GO) surface, its aromatic pi system is perturbed, which allow to control its electronic properties. In particular, the development of a band gap by means of chemical doping is a potent way to prepare G for use in nanoelectronic devices [14]. This type of functionalization includes two main routes: covalent bonding between free radicals or dienophiles and the C=C bonds of G and bonds between organic functional groups and the surface oxygenated moieties of GO. The covalent functionalization includes amidation, esterification, and introduction of negatively charged carboxylic or sulfonic groups. However, it typically introduces defects on the G network that have detrimental effect on the properties, in particular the electrical and mechanical.

The non-covalent functionalization is an interesting method since provides the possibility of anchoring functional groups to G without disturbing the electronic network. It involves the use of surfactants and/or polymers that wrap around G to facilitate dispersion via hydrophobic, π - π and ionic interactions [15]. It is believed that the hydrophobic regions cover the majority of the surface of the G sheets, whilst the hydrophilic tails stretch out into the aqueous solution where the flakes are dispersed [16]. Hence, the size and geometry of the resulting surfactant-wrapped G will significantly differ from that of pristine G.

An interesting approach is the isocyanate functionalization proposed by Stankovich *et al.* [17] that facilitates GO dispersion in polar aprotic solvents such as dimethylformamide (DMF), N-methylpyrrolidone (NMP) and dimethyl sulfoxide (DMSO). However, isocyanate-functionalized GO cannot be dispersed in non-polar solvents, and this restricts its use in organic polymers-based BHJ solar cells

In this regard, the present work focuses on the synthesis of hexamethylene diisocyanate (HDI)-functionalized graphene oxide. This nanomaterial possesses a highly hydrophobic nature and can be suspended in organic non-polar solvents such as benzene, toluene, chlorobenzene, tetrahydrofuran (THF), etc. Therefore, theHDI-GO sheets can be dispersed with organic materials, in particular conjugated polymers used for solar cell applications.

2 Experimental Section

2.1 Reagents

Natural graphite was obtained from Bay Carbon, Inc. (Michigan, USA). H_2SO_4, $KMnO_4$, P_2O_5, $K_2S_2O_8$, H_2O_2 (30 wt % in water), toluene, triethylamine (TEA) and hexamethylene diisocyanate (HDI) were purchased from Sigma-Aldrich and used as received. Dichloromethane (DCM) (analytical grade), acetone (99.9%), hexane (96%), pentane (99%), Tetrahydrofuran (THF) (HPLC grade); Dimethyl sulfoxide (DMSO) (99.5% anhydrous,); N,N-dimethylformamide (DMF) (99.8% anhydrous); N-methyl-2-pyrrolidone (NMP) (99%), methanol (99.8%), ethanol (analytical grade) and 2-propanol (99.5%) were purchased from Scharlau S.L. The aqueous solutions were prepared using ultrapure water obtained from a Milli-Q system (Millipore, Milford, USA).

2.2 Synthesis of graphene oxide (GO)

GO was prepared using a modified Hummers′ method from flake graphite [11,18]. Briefly, graphite powder, H_2SO_4,$K_2S_2O_8$, and P_2O_5, were heated at 80 °C for 5 h. After cooling, deionized water was added to the mixture and it was stirred overnight. The resulting product was then filtered, dried under air and oxidized again via addition of H_2SO_4, $KMnO_4$ and water in an ice-water bath. Following to dilution with water, excess $KMnO_4$ was decomposed by addition of 30 wt % H_2O_2 aqueous solution and then 5 wt % HCl aqueous solution.

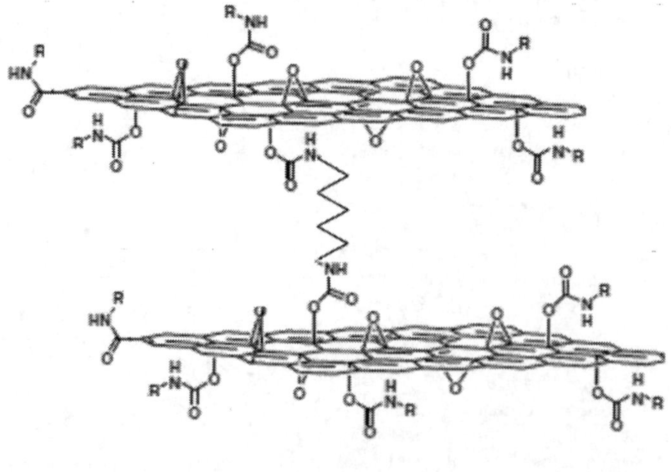

Figure 2. Schematic representation of HDI-functionalized GO

 https://doi.org/10.1051/e3sconf/20185702005

The product was filtered again and purified by repeating the following cycle: centrifugation, removal of the supernatant liquid, addition of aqueous solution of H_2SO_4 (3 wt %)/H_2O_2 (0.5 wt %) and bath ultrasonication for 30 min at a power of 140 W. Then it was washed several times with deionized water and finally freeze-dried.

2.3 Synthesis of hexamethylenediisocyanate-functionalized graphene oxide (HDI-GO)

1 g of purified GO powder was added to 100 mL of toluene, ultrasonicated in an ultrasonic bath for 2 h, and then transferred to a reactor equipped with mechanical agitator, thermometer and reflux condenser. Then 35 mL of TEA and 20 mL of HDI were were added to the system, the last one dropwisely. The solution was heated to 60 °C and vigorously stirred overnight under an inert atmosphere of argon. The mixture was then poured into methylene chloride to coagulate the product. Finally, the reaction product was filtered, washed thoroughly with methylene chloride and dried under vacuum to yield HDI-GO (Figure 2).

2.4 Instrumentation

An ultrasonic bath (Selecta, Spain) and a Hielscher UP400S ultrasonic probe (Hielscher Ultrasonics GmbH, Germany) equipped with a titanium sonotrode (Ø = 7 mm, l = 100 mm) was used to prepare the nanomaterial dispersions in the different solvents. Elemental analysis was carried out with a LECO CHNS-932 elemental analyser. Infra-red spectrum was obtained with a Perkin Elmer Frontier FT-IR Spectrometer.

3. Results and discussion

Similar to pristine GO, the HDI-functionalizedGO consisted in a black powder. However, unlike GO, it was insoluble in water or other polar solvents including DMF, NMP, or DMSO. This is in contrast to the results reported for GO functionalized by monoisocyanates, which were found to be soluble in polar aprotic solvents [17]. The insolubility of HDI-GO is likely ascribed to its higher hydrophobic character and the crosslinking between the GO sheets that are randomly covalently bonded by HDI to each other. Conversely, the synthesized nanomaterial was found to be soluble in organic non-polar solvents like benzene, toluene, chlorobenzene, THF, acetone, pentane, hexane, heptane and cyclohexane.

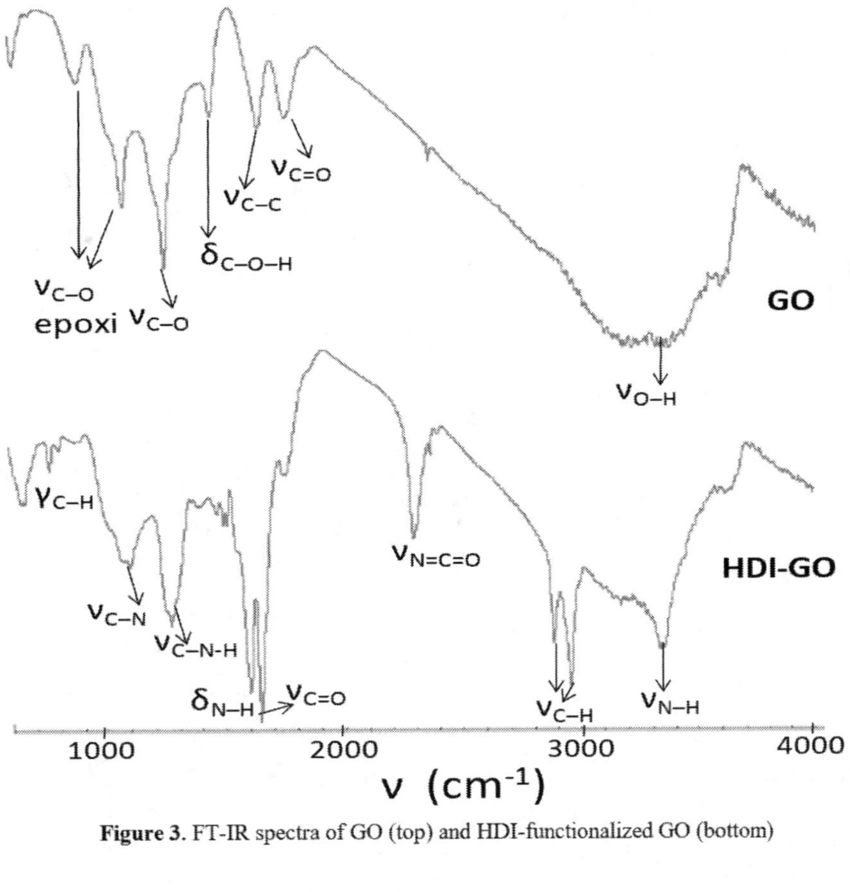

Figure 3. FT-IR spectra of GO (top) and HDI-functionalized GO (bottom)

The treatment of GO with isocyanates can lead to the derivatization of both the edge carboxyl and surface hydroxyl functional groups via formation of amides or carbamate esters [17], respectively. The chemical changes that occurred upon treatment of GO with HDI was investigated by Fourier transform infrared spectroscopy (FT-IR), since both GO and its derivative display characteristic IR spectra (Figure 3).

GO contains epoxide and hydroxyl functional groups on both sides of its basal plane and carboxyl moieties at the edge sites. The strong broad band centred at ~3520 cm^{-1} corresponds to the O–H stretching vibrations and that near 1750 cm^{-1} corresponds to the C=O stretching of the carboxylic groups. The band at 1600 cm^{-1} can be assigned to the aromatic C=C stretching, that at ~1400 cm^{-1} corresponds to the O–H deformation and those at approximately 1260, 1050 and 850 cm^{-1} are related to epoxy C–O stretching vibrations [18].

Upon treatment with HDI, the intensity of the O-H stretching band was strongly reduced, and a band appeared at 2280 cm^{-1} ascribed to the isocyanate group [19]. Further, a peak appeared at 3320 cm^{-1} due to N-H stretching vibrations, and those at 2850 and 2950 cm^{-1} arise from C-H stretching vibrations of the methylene chain of HDI. Intense bands are found at 1630 and 1585 cm^{-1}, attributed to the C=O stretching and N-H bending vibrations of the urethane or carbamate ester group [19]. The band ascribed to the O-H bending and those attributed to epoxy C–O stretching vibrations almost disappeared, indicating that the isocyanate reacted with both hydroxyl and epoxide groupsof the surface of GO to form carbamate esters. Further, the band close to 1110 cm^{-1} is likely related to -C(=O)-O and C-N stretchingvibrations of the urethane groups, and that at around 720 cm^{-1} is ascribed to the rocking of the methylene groups of HDI. Overall, the FT-IR spectra confirm the successful reaction of GO with the organic diisocyanate.

Assuming that the formation of carbamate esters through reaction of HDI with the OH or epoxide groups of GO is the hydroxyls or epoxies is the unique reaction pathway, the nitrogen-to-carbon atomic ratio can be used to roughly estimate the functionalization degree (FD). In particular, FD can be expressed as moles of carbamate ester unit incorporated per mol of carbon atoms of GO. Taking into account the results from elemental analysis of GO: C, 41.93%; O, 51.96%; H, 3.44%; S, 2.67% and HDI-GO: C, 56.68%; O, 28.22%; H, 5.82%; N, 8.00; S, 1.28%, the calculated FD was 16.32%, higher than that reported for other isocyante functionalized GO [17]. It should be noted that this value corresponds to a lower bound of functionalization; in the case of amide formation, which seems less feasible according to the FTIR spectrum, the FG would be higher since there is a loss of carbon (as carbon dioxide) from the isocyanate reagent.

4 Conclusions

Hexamethylene diisocyante-functionalized graphene oxide (HDI-GO) has been prepared following a two-step approach: firstly, GO was prepared using a modified Hummers´ method from graphite, and secondly GO was treated with HDI in toluene to yield the modified nanomaterial. HDI-GO was found to be soluble in organic non-polar solvents like benzene, toluene, chlorobenzene, THF, acetone, pentane, hexane, heptane and cyclohexane. The comparison of the elemental analysis and FT-IR spectra of GO and HDI-GO corroborated the successful synthesis of the functionalized GO. Future work will focus on the characterization of HDI-GO by UV–vis spectroscopy, X-ray diffraction, atomic force microscopy, scanning electron microscopy, transmission electron microscopy and Raman spectroscopy. Further, mixture of these derivatives with different types of conductive polymers will be tested, with the aim to use them in photovoltaic cells.

Acknowledgments

Financial support from Fundación Iberdrola España via an Energy Research Grant 2017 is gratefully acknowledged. We would also like to thank to Alcalá University for making possible this research.

References

1. Geim A K and Novoselov K S 2007 *Nat. Mater***6** 183
2. Novoselov KS, Geim AK, Morozov SV, Jiang D, Zhang Y and Dubonos S 2004 *Science***306** 666
3. Lee C, Wei X, Kysar J W and Hone J 2008 *Science***321** 385
4. Novoselov K S, Falko V I, Colombo L, Gellert P R, Schwab M G and Kim K 2012 *Nature* **490** 192
5. Dubacheva G V, Liang C K and Bassani D M 2012 *Coord.* Chem. *Rev.***256** 2628
6. Iwan A and Chuchmała A 2012 Prog.Polym. Sci. 37 1805
7. Manzano-Ramirez A, Lopez Naranjo E J, Soboyejo W, Meas-Vong Y and Vilquin B 2015 *J. nanomaterials* **2015** 406597
8. Xiao-Feng L, Zi-Yan Z, Zhong-Ke Y, Jing L, Xiao-Fen X, Wei H, Xu-Dong C and Ding-Shan Y 2016 *Chinese Chem. Lett.* **27** 1259
9. Park S J and Ruoff R S 2009 *Nature Nanotechnol.* **4** 217
10. Steurer P, Wissert, Thomann R and Mulhaupt R 2009 *Macromol. Rapid Comm.* **30** *316*
11. Hummers W S and Offeman R E 1958 *J. Am. Chem. Soc.* **80** 1339
12. Díez-Pascual A M, Gómez-Fatou M A, Ania F and Flores A 2015 *Prog. Mater. Sci.* **67** 1
13. Texter J 2014 *Curr. Opin. Coll. Interf. Sci.* **19** 163
14. Niyogi S, Bekyarova E, Itkis M E, Zhang H, Shepperd K, Hicks J, Sprinkle M, Berger C, Ning Lau C, de Heer W A, Conrad E H and Haddon R. C. 2010 *Nano Lett.* **10** 4061

https://doi.org/10.1051/e3sconf/20185702005

15. Díez-Pascual A M, Hermosa C, San Andrés M P, Valiente M and Vera S 2017 *J. Phys. Chem. C* **121** 830
16. Mateos R, Vera S, Valiente M, Díez-Pascual A M, and San Andrés M. P. 2017 *Nanomaterials* **7** 403.
17. Stankovich S, Piner R D, Nguyen S T and Ruoff R S 2006 *Carbon* **44** 3342
18. Diez-Pascual A M and Diez-Vicente A L 2016 *ACS Appl Mater Interfaces***8** 17902
19. Silverstein R M, Webster F X and Kiemle D J 2005 *Spectrometric Identification of Organic Compounds*, 7th edition (John Wiley & Sons, Inc, Hoboken, USA) p. 72-126

SOC Estimation of Lithium-Ion Battery Based on Kalman Filter Algorithm for Energy Storage System in Microgrids

Dae-Won Chung , *Seung-Hak* Yang

Dep't of Electrical Engineering, Honam University, Gwangju-city, South Korea

Abstract—State-of-charge (SOC) is one of the vital factors for the energy storage system (ESS) in the microgrid power systems to guarantee that a battery system is operating in a safe and reliable manner for the system. Many uncertainties and noises, such as nonlinearities in the internal states of a battery, sensor measurement accuracy and bias, temperature effects, calibration errors, and sensor failures, pose a challenge to the accurate estimation of SOC in most applications. This study makes two contributions to the existing literatures. First, a more accurate extended Kalman filter (EKF) algorithm is proposed to estimate the battery nonlinear dynamics. Due to its discrete form and ease of implementation, this straightforward approach could be more suitable for real applications on the ESS. Second, its order selection principle and parameter identification method are illustrated in detail. It can accurately demonstrate the characteristics of the lithium-ion battery to show the feasibility and effectiveness of the algorithm for the ESS.

1 Introduction

One of the most remarkable differences between traditional energy production and distribution systems and present ones is that most of the power generation is decentralized to smaller plants located near the end-use points called microgrids. A microgrid is a localized physical space that consists of distributed power generation, storage, and consumption. The need for achieving a balance between the electric power generation and the demand of consumers has added to the emergence of smaller power generation and demand environments, in which the adaptation of production to load can be performed much more dynamically due to their distributed smaller elements and the geographical proximity of all elements in the microgrids.

The most important techniques is the energy storage components for using energy later, and the energy storage system (ESS) on the grid for various purposes, such as grid stability, peak shaving, and renewable energy time shifting. The batteries play a significant role in the microgrids. Due to the volatility, flammability and entropy changes, batteries have a potential to ignite or even explode when overcharged. The solutions to all these problems lie not only in battery technologies, but also the management systems for the batteries. The battery management system (BMS), which integrates the battery system with the rest of the power train, plays a vital role in improving the battery performance, safety, and reliability. Meanwhile, the accurate estimation of state-of-charge (SOC) is one of the main tasks of the BMS, which will help improve the system performance, reliability, and safety, and will also lengthen the lifetime of the battery. In fact, the precise SOC estimation of the battery can avoid unpredicted system interruptions and prevent the batteries from being over charged and over discharged, which may cause permanent damage to the internal structure of the batteries [1].

The SOC means the proportion of released or stored energy from all the battery energy storage. Only when the SOC is estimated precisely, the energy balance can be achieved without damaging the batteries. However, due to a battery's complex internal chemical and physical reactions, the SOC cannot be measured directly and multiple factors should be considered. Thus, it is difficult to estimate the SOC when the battery is working.

The general approach for measuring the SOC is to indirectly estimate both the coulombs and current flowing in and out of the cell stack under all operating conditions and the individual cell voltages of each cell in the stack accurately [2]. The coulomb counting (CC) method, which measures the current flowing into and out of the cell, is the most commonly used estimation method [3], [4]. However, the integral calculation of the current flow has an accumulation error; with the increase in the operating time, the accumulation error will result in a larger SOC estimation error. The other normally used SOC estimation method includes the open-circuit voltage (OCV) method, neural network method, and so on. The OCV method is often used in the battery overhaul rather than the SOC estimation; neural network models are too complicated to train the physical model and to need a plenty of operating data [5].

The method presented in this study is the extended algorithm based on the Kalman filter, which is used to estimate the inner states of any dynamic system. The core of the extended Kalman filter (EKF) algorithm is to propose a precise state describe model, according to the

https://doi.org/10.1051/e3sconf/20185702006

measurement data, to calculate the values of the objects with the recursive algorithm. Comparing to other algorithms, the EKF algorithm has its own advantages: even if there were errors in the system, self-correction can be down to a certain range and the optimal estimation value can be obtained. The EKF algorithm needs accurate state describe model and accurate measurement values.

2 Modeling of lithium-ion battery

The purpose of building a battery model is to establish the battery's internal state variables and external quantitative relation electrical characteristics, and to set up the mathematical model. Further, based on the external variables, such as battery voltage, current, and temperature, a battery model helps to calculate the internal state variables, such as SOC, internal resistance, and electromotive force. Many types of battery models were proposed to guarantee the precision in the previous literatures [2,4]. An equivalent circuit model [5] of them to represent schematically a chemical reaction inside a lithium-ion battery pack is shown in Fig. 1. The bulk capacitance (C_b) represents the battery pack storage capacity, and the surface capacitance (C_d) represents the battery diffusion effects to describe the cell's electrochemical polarization and concentration polarization, which reflects the battery transient response of charge or discharge. The resistances R_i and R_d represent the internal resistance and the polarization resistance, respectively. The voltages across the bulk capacitor and the surface capacitor are denoted by V_b and V_d, respectively. The battery pack terminal voltage and the terminal current are denoted by V_o and I, respectively. The parameters required for the battery model can be determined from the experimental data, where the OCV tests are performed upon the successive discharge of a battery by injecting the current pulses as described in section Ⅳ.

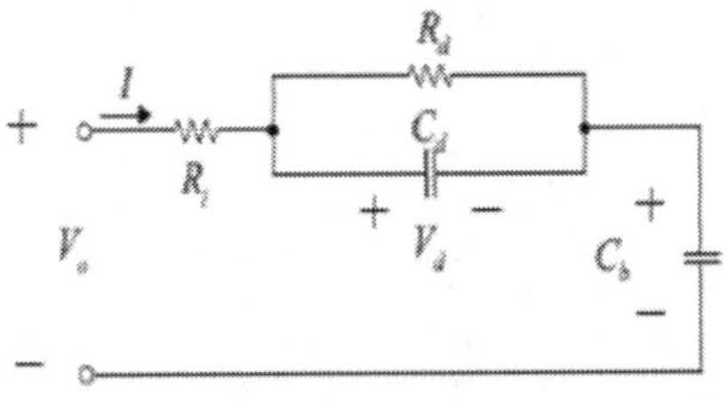

Fig. 1. Equivalent circuit model for a lithium-ion battery pack

This kind of model has a good dynamic performance and can simulate the dynamic characteristics of the lithium-ion battery accurately [6].

According to the Kirchhoff's voltage and current laws, the capacitance voltage variation, and its current relationship, the characteristics of the model in Fig. 1 [4] can be expressed as follows:

$$\dot{V}_b = \frac{I}{C_b}, \tag{1}$$

$$\dot{V}_d = -\frac{I}{R_d C_d} V_d + \frac{1}{C_d} I, \tag{2}$$

$$V_o = V_b + V_d + IR_i. \tag{3}$$

Since the relationship of OCV and SOC for the lithium-ion battery is only piecewise linear in practice, the voltages across the bulk capacitor can be expressed as

$$V_b = k * SOC + d, \tag{4}$$

where the k and d coefficients are not constant and vary with the battery SOC and ambient temperature. Thus, they can be expressed as

$$\begin{bmatrix} SOC \\ V_d \end{bmatrix} = \begin{bmatrix} \dfrac{I}{kC_b} \\ -\dfrac{I}{R_d C_d} V_d + \dfrac{1}{C_d} I \end{bmatrix}, \tag{5}$$

$$V_o = k * SOC + V_d + IR_i + d. \tag{6}$$

These two equations (5) and (6) are indicated as a model of the state and observation equations for the lithium-ion battery pack.

The curve graph of OCV based on the nonlinearity of the SOC can be obtained in Fig. 2.

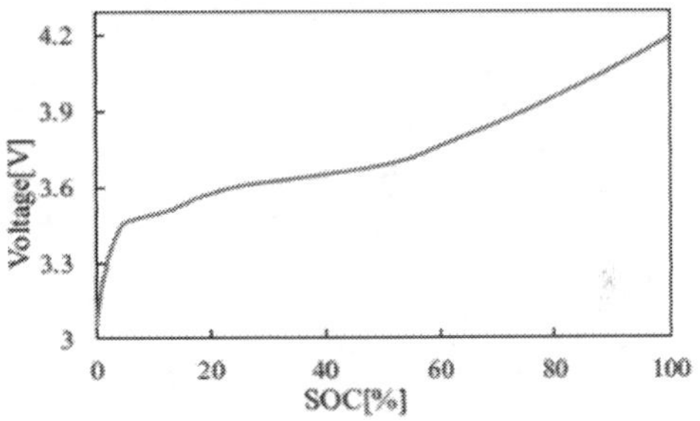

Fig. 2. OCV versus SOC curve graph.

3 SOC estimation principle of EKF

In case that the system's state equation is nonlinear, the Kalman filtering method cannot be used directly, i.e., the linearization system equation by using the Taylor series expansion is required; this type of linear equation for the nonlinear model of the Kalman filtering method is called the EKF method [4].

According to (5) and (6), the system state equation can be written as

$$X = [x_1 \quad x_2]^T, \; x_1 = SOC, \; x_2 = V_d, \tag{7}$$

$$u(t) = I, \; y(t) = V_o. \tag{8}$$

If the system input of a model is defined as u(t) = I and the output is defined as y(t) = V_o, then the battery model can be written as

$$\dot{x} = f(x,u) + w, \tag{9}$$

$$y = g(x,u) + v. \tag{10}$$

where,

$$f(x,u) = \begin{bmatrix} \frac{u}{kC_b} \\ -\frac{I}{R_d C_d}x_2 + \frac{1}{C_d}u \end{bmatrix}, \tag{11}$$

$$g(x,u) = k x_1 + x_2 + R_i u + d. \tag{12}$$

In these formulas, w and v represent the system noise and measurement noise, respectively. For (9) and (10) of the nonlinear model, the model after linearization by using the Taylor series expansion is obtained.

$$\delta\dot{x} = A_t \delta x + B_t \delta u + w, \tag{13}$$

$$\delta y = C_t \delta x + D_t \delta u + v, \tag{14}$$

where

$$A_t = \frac{\partial f(x,u)}{\partial x}\bigg| x(t), u(t) = \begin{bmatrix} 0 & 0 \\ 0 & -\frac{I}{R_d C_d} \end{bmatrix}, \tag{15}$$

$$B_t = \frac{\partial f(x,u)}{\partial u}\bigg| x(t), u(t) = \begin{bmatrix} \frac{1}{kC_b} \\ \frac{1}{C_d} \end{bmatrix}, \tag{16}$$

$$C_t = \frac{\partial g(x,u)}{\partial x}\big| x(t), u(t) = \begin{bmatrix} k & 1 \end{bmatrix}, \tag{17}$$

$$D_t = \frac{\partial g(x,u)}{\partial u}\big| x(t), u(t) = R_i. \tag{18}$$

For the digital computation of the linearized model of (13) and (14), the system equations for the continuous time equation should be expressed by the discretized time model, as represented in (19) and (20).

$$x_{k+1} = A_d x_k + B_d u_k + w_k. \tag{19}$$

$$y_k = C_d x_k + D_d u_k + v_k \tag{20}$$

where the parameters in both the equations are,

$$A_d = I + A_t T_s, \tag{21}$$

$$B_d = B_t T_s, \tag{22}$$

$$C_d = C_t, \tag{23}$$

$$D_d = D_t, \tag{24}$$

where I is the unit matrix and T_s is the sampling period, respectively.

As per the above equations, the Kalman filter is an optimal observer whose principle is illustrated in Fig. 3. The principle is to minimize the errors between the estimated and the measured outputs in real time, using a feedback that adjusts the uncertain variables of the used model [7]. By such a model, it is possible to observe the physical parameters of the model that are not accessible to measurements. The correction is weighted by a gain vector K that allows the correction of the dynamics and the performance of the filter. The gain is calculated at each iteration from the error predictions and uncertainties (noise) on the states and measurements. The filter dynamic control is then based on the initialization of the noise matrices of states Q and measurements R, as well as through the initialization of the matrix of error covariance P.

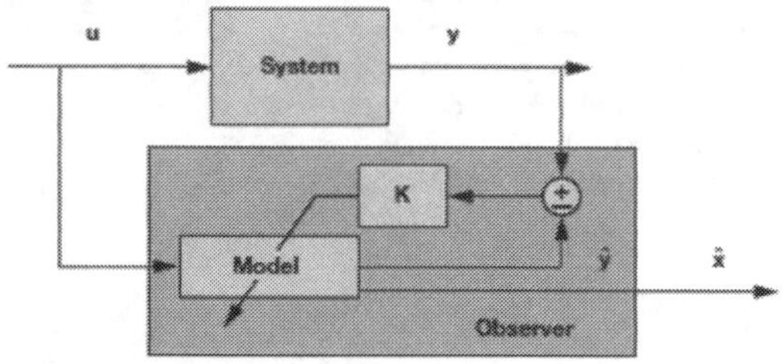

Fig. 3. Principle block diagram of the Kalman filter.

To realize the optimal observer as indicated in Fig. 3, the Kalman filter algorithm can be divided into two parts [7] as follows:

- Predict system status, system output, and error
- Correct the current state estimate value based on the system output value

This takes place in two phases: the first phase has the initialization of the matrices P, Q, and R; and the second phase includes observation, which is composed of two steps at each sampling interval [8], as shown in Fig. 4. First, the algorithm predicts the value of the present state, output, and error covariance. Second, by using a measurement of the physical system output, it corrects the state estimation and error covariance. The core of EKF was to compare the predicted value and the measurement value, adjusting the Kalman gain. According to the size of the error, the gain will be used to calculate the next predicted value. The bigger the error, the greater the gain, and the bigger the estimated value will be corrected. Similarly, the lesser the error, the smaller the gain, the lesser the estimated value will be corrected. So, the EKF algorithm is applied to the SOC estimation for a lithium-ion battery pack.

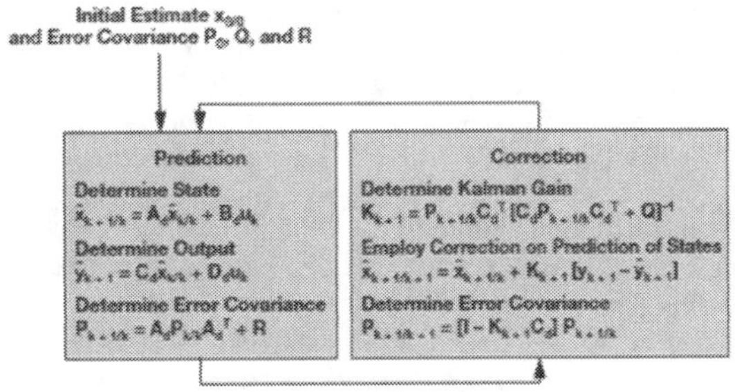

Fig. 4. Computation steps of the EKF algorithm.

4 Simulation and experiment

4.1 Model parameter extraction

All the parameters proposed in the model are the functions of SOC, current, and temperature. However, only the dependency of SOC and current is analyzed because of the aforementioned reasons. The previous researches have used electrochemical impedance spectroscopy (EIS) for extracting the parameter values. However, this technique cannot be utilized in practical applications, where the operation is much more dynamic and unpredictable. Another much simpler theory [9] behind the parameter extraction involves Ohm's law, which relates the proportionality between the instantaneous voltage and current. However, this technique works well only if the time step used is smaller than the time constant of the associated reaction, so that the change in the state can be considered minimal. A much more advanced technique includes the application of the curve fitting tools [10] in MATLAB to identify the values of R_i, R_d, C_d, and C_b. The value of R_i is determined using the instantaneous change in the value of the voltage at the end of a discharge pulse. The technique proposed in this study is similar to the one as presented in [11]. The difference lies in the fact that the model in [11] does not consider the effect of the varying levels of the discharge currents. However, the proposed model in this study includes this effect.

4.1.1 Pulse discharge test

The pulse discharge test (PDT) includes the discharging of the battery pack by providing the current in the form of constant pulses. There is a rest time included between the application of two subsequent pulses as shown in Fig. 5. The rest time is decided based on the settling of the OCV to steady state. The discharge pulses are supplied to the pack to reduce the SOC by 15% with each pulse. The duration of each pulse is calculated using the SOC dependency equation on the discharge current. The main aim of these tests is to extract the model parameters, R_i, R_d, C_d, and C_b. The PDTs are performed at varying pulse current levels to identify the effect of the discharge current on the model parameters. These tests are also used to determine the OCV-SOC curves. The discharge current values are decided based on the maximum current rating of the cell specified in the data sheet.

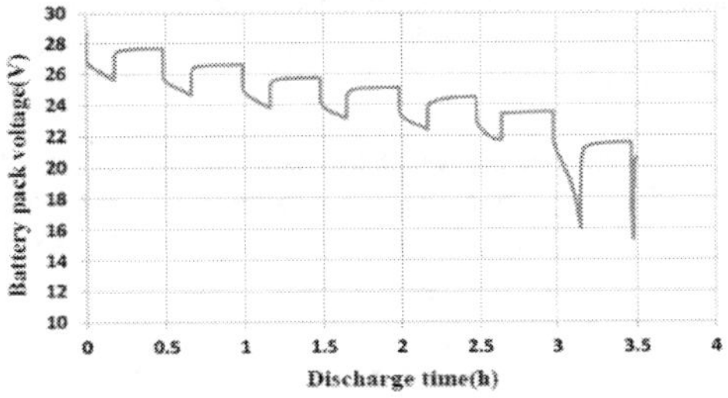

Fig. 5. Pulse discharge voltage profile of the battery cells.

4.1.2 Continuous discharge test

The continuous discharge test (CDT) involves the discharging of a battery pack at constant continuous current levels as shown in Fig. 6. The current levels remain the same as for the PDTs. These tests are used to identify the effect of the different discharge currents on the capacity of the battery pack and of each cell.

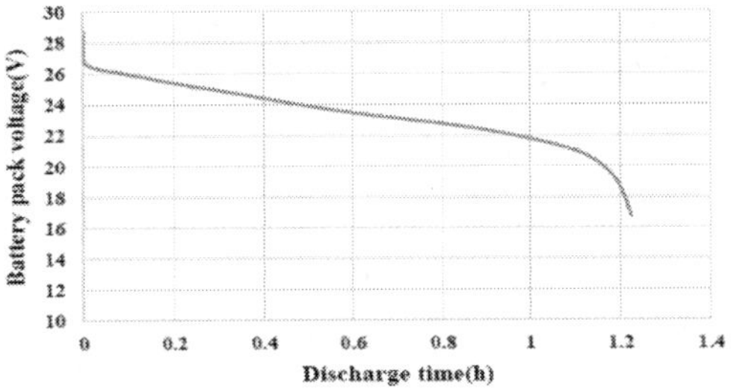

Fig. 6. Continuous discharge voltage profile of the battery cells

4.2 Experimental results

To simulate lithium-ion battery charge and discharge state in a normal use, and verify the effectiveness of the SOC estimation procedure, the following charging experiments have been conducted. Under a normal temperature, the battery is charged from OCV = 3 V reference state (namely, SOC 0% status) under a constant current until OCV = 4.2 V (namely, SOC 100% status). The state change, the error of the voltage, and the SOC waveform diagram are observed. The SOC estimation results and the expected integral values are basically the same, which explains that the EKF algorithm can achieve the accurate SOC estimation values. We evaluated the accuracy of the SOC estimation. The current load applied to the target lithium-ion battery is shown in Fig. 7. Charge and discharge produces the current load pattern shown in Fig. 8. This charging method is the constant current-constant voltage (CC-CV) charging topology. The experiments are discharged on the above current patterns until the battery SOC reaches 0.0%. The experiments are evaluated by comparing the estimated SOC by EKF that implements the MCU microcomputer and SOC calculated by the charging and discharging

 https://doi.org/10.1051/e3sconf/20185702006

machine. The SOC estimation accurately evaluates the absolute value by subtracting the true SOC from the estimated SOC in percentage. The true SOC is given by the SOC value, which is given by the highly accurate experimental equipment. The time transition of the estimated SOC and reference SOC is shown in Fig. 10. The average error for the SOC estimation experiment is 0.15%. The experimental results show that the proposed EKF-based SOC estimation method is effective and can estimate the battery SOC accurately.

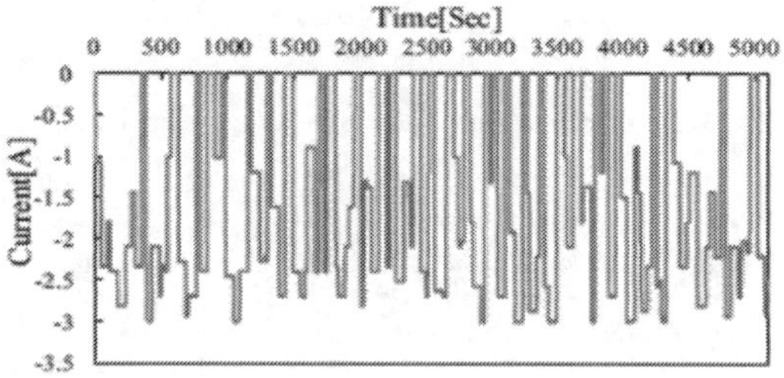

Fig. 7. Discharge current load.

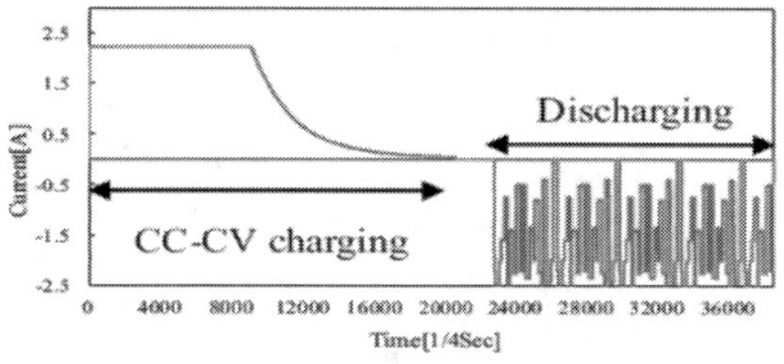

Fig. 8. Charging and discharging current method.

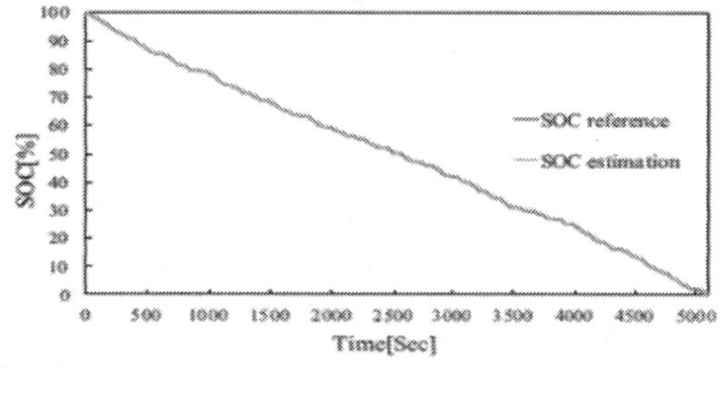

Fig. 9. SOC estimation results.

5 Conclusion

In this study, a new method for the SOC estimation based on the EKF algorithm has been presented for the ESS in the microgrids, after obtaining the relationship of the lithium-ion battery SOC and OCV. First, the parameters of the lithium-ion battery model are confirmed, and then these parameters are plugged into the state and observation equations. The measured current and voltage are used in equations. Further, through the iterative calculation of the EKF equations and constant updating of the system status, an accurate estimation value can be obtained ultimately. This type of EKF calculation algorithm has been validated through the simulation and experiments. Thus, the SOC estimation based on the EKF algorithm is effective and accurate. If the difference between the noise value in equations and the actual noise value is substantial, the estimated value is difficult to converge to a reasonable range. Estimation of the influence of noise in the algorithm is the next step in the research direction.

This method may be applied for an accurate estimation of SOC to improve the ESS performance, reliability, and safety, and to lengthen the lifetime of the battery on ESS in the microgrid.

References

1. P. Bentley, B.S. Bhangu, C.M. Bingham, and D.A. Stone, "Nonlinear observers for predicting state-of-charge and state-of-health of lead-acid batteries for hybrid-electric vehicles," *IEEE Trans. Veh. Technol.*, **54**, 783–794 (2005)
2. M. Verbrugge and B. Koch, "Generalized recursive algorithm for adaptive multiparameter regression," *J. Electrochem. Soc.*, **153**, A187–A201 (2006)
3. K.S. Ng, C. Moo, U. Chen, and Y. Hsieh, "Enhanced coulomb counting method for estimating state-of-charge and state-of-health of lithium-ion batteries," Applied Energy, **86**, 1506–1511 (2009)
4. F. Zhang, G. Liu., and L. Fang, "A battery state of charge estimation method with extended Kalman filter," IEEE/ASME International Conference on AIM, (2008)
5. J. Andreas, M. Perrin, and S. Piller. "Methods for state-of-charge determination and their applications," *J. Power Sources*, 96, 113–120 (2001)
6. M. Kitagawa, H. Shibuya, and J. Takehara, "Development of battery state of charge indicator for electric vehicles," Proc. of the 11th. International Electric Vehicle Symposium (1992)
7. N. Kawarabayashi, L. Lin, R. Ishizaki, and M. Fukui, "A battery smart sensor and its SOC estimation function for assembled lithium-ion batteries," SASIMI 2015 Proceedings. (2015)
8. M. Murnane, Solar PV Systems, Analog Devices, Inc. Adel Ghazel Chief Technology Officer, EBSYS Technology Inc./WEVIOO Group, "A closer look at state of charge (SOC) and state of health (SOH) estimation techniques for batteries," Technical Article, www.analog.com
9. T. Huria, M. Ceraolo, J. Gazzarri and R. Jackey, "High fidelity electrical model with thermal dependence for characterization and simulation of high power lithium battery cells," Electric Vehicle Conference (IEVC), 2012 IEEE International, Greenville, SC, 1-8 (2012)
10. MATLAB user manual, https://www.mathworks.com/
11. M. Dubarry, N. Vuillaume, and B.Y. Liaw, "From single cell model to battery pack simulation for Li-

https://doi.org/10.1051/e3sconf/20185702006

ion batteries," *J. Power Sources*, **186**, 505–507 (2009)

Research and Discuss on Market model based on Blockchain

YAN Cui-hui[1] , *LIU* Fang[1], *HU* Chaofan[2], *CUI* Hui[1] and *WANG* Bin[2]

[1] Power System Automation Department, China Electric Power Research Institute, Haidian District, Beijing
[2] National Power Dispatching and Control Center, State Grid Corporation of China, Beijing, China

Abstract. Electric power industry is the foundation of national economy and plays a key role in the development of human society. Traditional power industry is monopolized by industry and lacked of market efficiency. In recent years, it is seeking breakthroughs in the market-oriented reform. Although China's electricity market reform has achieved some achievement, it is still lacked of market players, regional transactions, excessive data management, concentration and other defects. This paper puts forward higher requirements for the system design of power market transactions. The electrical power market is facing with the security and performance challenges, it is to improve the electrical power market transaction mode and to enhance transaction security, blockchain is a new decentralized framework and with a distributed computing paradigm, which integrates and innovates a variety of computer technologies, then it constructed a perfect competitive electricity market system. Using of different levels of multi auction trading model to integrate the province's trading and inter provincial trading parts in one system. In the part of inter-province trading, it adopted the contract trading and bidding coexist pattern, the trade is carried out in each province Surplus-deficit electricity which based on the meet of the province's demand basis. By using blockchain, it can conducted the intelligent contract and record the whole transaction information discretely, and provided the inquiry transaction channel to the users, so it can achieve the resource scheduling fairness and transaction information in security.

1 Introduction

Electrical power market reform has initiated in China since 2000, and it have made experiments in several provinces, it realized the separation of plant and network and achieved the competitive bidding construction of sell side market as in refers [1], the electricity market trading volume increased rapidly over the past two years, the total electricity consumption has gone up to 20% through market transactions in 2016 [2]. However, there are still some problems in the electricity market, for example, the power trading is mainly concentrated of the province market and the management mode is over centralization [3]. On the one hand, its results in the existing electricity market is difficult to meet the market demand, on the other hand, the centralized data management leaded to the low data security and credibility, which also prejudice the development of market integration and resource allocation [4]. In order to solve the above shortcomings, it is necessary to seek a decentralization platform technology to support large-scale regional electricity market transactions and more secure and reliable transaction data storage. Using the new emerging decentralization technology of blockchain, the problems in current power market transactions can be much better solved.

2 BlockChain

2.1 Brief Introduction of electricity market

Electricity market is a place or platform where the buyer and seller of electricity interact with each other, then to determine the clearing price and quantity, it has competitive and open characters, then mainly configure the power resources using market measure. As in refers [5] it is a new type of power operation mode based on fair competition among electrical power enterprises, which can effectively reduce power production cost and raise production efficiency.

The existing electricity trading patterns differences mainly focused on whether Power Grid Corp involved in electricity transactions or responsible for power transmission only [6]. Most of the world countries adopted the free market model in early, which Power Grid Corp is only responsible for power transmission. In American model [7], the Power Grid Corp is not take part in the purchase and sale of electricity business, which only in charge of power grid management and wheeling charge [8]. Canada mainly used the power generation and sale link independent mode, it is monopolized of the Power Grid Corp operating in the transmission and distribution sectors, but in the end of power production bidding bringing in competition, then in the customers right brought in market competition and open the clientele choose independently[9]. India and Brazil mainly adopted the transmission and consumption link independent mode, Power Grid Corp did not participate in market transactions only introduced competition on the power generation side [10]. In the Commonwealth model, the national electrical power company possessed all the power transmission networks, and did not participating in the market of the power generation and the selling side, which fully introduced the market competition [11]. Northern Europe adopted centralized operation and market mode at the same time, where the power generation companies and power users can complete the transaction on the self-supporting power transmission and distribution network, and also can charge net service fees by the Power Grid Corp [12]. In the Argentina electricity market model, it has more realized the independence of all aspects of electricity market links thoroughly such as the generation, transmission, distribution and selling, the Power Grid Corp has completely not participated in the electricity market transactions as in refers [10]. Only Japan considered energy security and its specific conditions, it still used integrated system of power supply, transmission and distribution as referred in [10] [13] [14].

China power industry reform has experienced the stages of complete monopoly, "collecting funds for electricity", simulating electric power market, electricity market pilot, "generation and network separation" reform and regional electricity market pilot operation stages in refers [15]. Since the 2015 electricity reform No. 9 release as in refers [16], the State Grid is implemented the market-oriented reform of the power grid, it aims to achieve a competitive electricity market, which can orderly liberalization the electricity price of competitive aspects and the power plan outside of public welfare adjustment, it can strengthen the electricity market platform and voluntary and increase the inter provincial and cross regional market transactions, then enhance market efficiency and promote the power resources configuration optimally in market regulation and in a larger scope, it can improve the network operation efficiency and power supply service level as in refers [17]. According to the policy reform and market construction of several pilot provinces, the typical power trade methods and trading power of several provinces is shown in table I [18].

Table 1. Typical provinces electricity markets trading methods and electricity

Province	*Transaction mode*	*2016year transaction power*	*electricity proportion consumed in this province*
Guang dong	Bilateral consultations (long association) and monthly centralized bidding	43 billion 980 million kwh	8.06%
Shan dong	Bilateral consultations, centralized bidding and listing Trading	129 billion 509 million kwh	24.02%
Shan xi	Off-site bilateral (multilateral) consultations, on-site bidding (including match), floor listing Trading	11 billion 90 million kwh	6.17%
An hui	Bilateral trading and centralized trading	39 billion 200 million kwh	21.84%
Si chuan	Autonomous bilateral trading and centralized trading	Not yet carry out the sale of electricity business	/
Jiang xi	Bilateral trading oriented, centralized trading as a supplement	15 billion kwh	12.69%

2.2 Problems in the current Electricity Market

Although the development plan described in the previous section has many advantages [19][20], the current pilot electricity market model still has defects in considerable degree. On one hand, the management mode is over centralized; on the other hand, the market body lacks the regional electricity market[21].

The current mainstream electricity trading is mainly based on the main center operation mode, which can be operated normally in the case of government unified pricing, it will lead to the management of electricity trading increasingly much more and the information out of sync after the introduction of market factors. First of all, the data amount and real-time demand produced in the electricity trading would increasing times, there would be needed large number of manpower to maintain the central data and the trading clearing process, but the manual operation caused the trading clearing process inefficiently and trade confirmation delay. Secondly, according to the development of distributed generation technology, it will cause the electric bilateral trade mode produced because of mass generator and consumers brought in, and the original model cannot support the demand as well. Finally, once the organization attacked, the data can be lose or tampered, the excessive centralization will lead to the information out of sync, user' privacy faced threat and may appeared the condition of using central authority to damage the participant interests. Therefore, we need to find a decentralized, dynamic transaction data, and a high degree of automation technology to support the operation of the new electricity market trading platform as in refers [22] [23].

From the point of market players, all kinds of pilot are taking the province as market main body. Considering of the market development and resources allocation, it would lay a hidden danger for the future of market development. From the point of the future development model, the proportion of market is still small although the electricity consumption between provinces increased. Taking the power industry overall situation of 2017 year in 1-3 for example, the electricity sent out of each provinces including regional market trading taking up to 14.9 proportion in country total electricity consumption, the cross-district delivery electricity only accounted for 5.12 proportion of total [24]. It can be assumed that the amount of transmission electricity coming from electricity market trade would much less than this value. Most of china's electricity market is still limited to inner provinces dispatching and inter-provinces dispatching which center is based on the State Grid and South China power Grid, it caused the weak connection between various regions, when the inter province market over centralized, it would lead to market rigidity easy and then bring about administrative interference and local protection problems, so it is difficult for the market to play a role in system regulatory, the operation efficiency is far less than the regional electricity market [5]. From the point of resources allocation, the energy distribution and load demand between each provinces has bigger differences, the supply and demand in province is imbalance. In the provincial market system solidified condition, it is easy to produce the provincial barriers, then can limited the trade volume between provinces resources and restricted the industrial development [25].

In conclusion, the problems exited in current pilot electricity market model is as follows:

First, the management methods is much more concentrated, it lead to the difficulty to trade conduct and data storage increased greatly. Meanwhile, the trade information is not open to the traders, it is decentralized and the data credibility is low. The database of central structure can easy to attack with no perceive.

Secondly, the electricity market is mainly concentrated in the province area, the operation efficiency is low, which is not facilitated for the regional market developed in a wider scope, it also restrict the resources allocation between each provinces and reject the development of economic industry.

2.3 Ethernet square Intelligent Contract

The technology mentioned in the previous section is supported the platform, with the characteristics of weak centralized, data security and cannot be tampered with, it is open and transparent to query, reliable, support disposing a large number of data and storage, it is with higher automatic execution and the management cost reduced, about all, it is considered to design system using Ethernet and intelligent contract. The relevant conceptions is giving as follows.

Blockchain is a new technology, which is developed gradually with the digital encryption currency bitcoin, it provides a credit establishment form with decentralized and dispensed with no credit accumulation paradigm [7] [26].

According to establish an all maintained and no tampered database to record all the trade recordings and historical data, all data are stored distributed and public. In essence, it is a bunch of data block linked together, each data block record a special set of trade status information with arborescence feature, then combined with distributed database and timestamp techniques, it can ensure the block safety and the trade data cannot be tampered. Under this technology, any unknown network users can reach an agreement according to contract, point to point accounting and digital encryption methods, which is completed without any central trust institutions [7] [27] [28].

The platform used in this paper is ethereum, which is the intelligent contract development platform based on blockchain. The ethereum include external accounts and contract accounts, among these, the contract accounts can be controlled by the code of stored accounts. When a message is sent from one account to another, there would be a trade executed and the code and input data in the target account will be executed too [29].

The smart contract writing with ethereum can realize self-conduct and validate once established and deployed, it does not require human intervention, and produced verifiable evidence to demonstrate the effectiveness of contract operation [30]. It can realize a contract which with clear logic, easy to test, clear responsibility and simple to find out responsibility. The malicious or unexpected incident would take place reducing to minimum, the demand of trust intermediary become low, and also reduce the loss caused by cheat [7].

3 Application feasibility of Blockchain in Schedule system

We adopt the mode of "contract plus bid" to build an integrated electricity market trading platform composited the regional market and the provincial electricity market.

It integrated different levels of trading within the same system, it can ensure the power supply and demand balance inter provinces in prior when signed contract volume, the provinces can share the Surplus-deficit electricity and bidding price, it can execute competitive price trade in region and achieve efficient market trading. Here, it adopted the smart tract based on ethereum, the trade process is executed automatically according to the established rules, the labor cost in system maintenance can be reduced, and the credibility of trade has also increased. The whole trade data system is decentralized, the trade information can be searched by both sides of trade at any time, and the data cannot be tampered, then it can ensure the trade fair, justice and open.

In summary, we constructed the safe and efficient electric power market trading system, it can realize the trade data out of tampering and inquire safely which the electricity can allocated fairly in large scale region market, improve the production efficiency and accelerated the development of electricity industry.

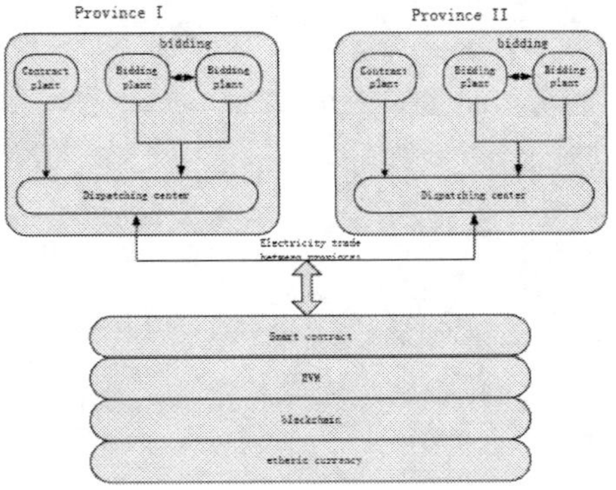

Fig. 1. Actual system framework.

This work used the Solidity language to write smart contracts and uses it as a technical support to build an impeccable competitive cross regional electricity market system. Among them, the smart contract is a deployed on blockchain and can run on ethereum machine. The competitive regional electricity market system constructed in this paper can realize the intra provincial contract electricity, bidding electricity trading and inter provincial electricity trading in the electricity market. The actual system diagram is shown in figure 1.

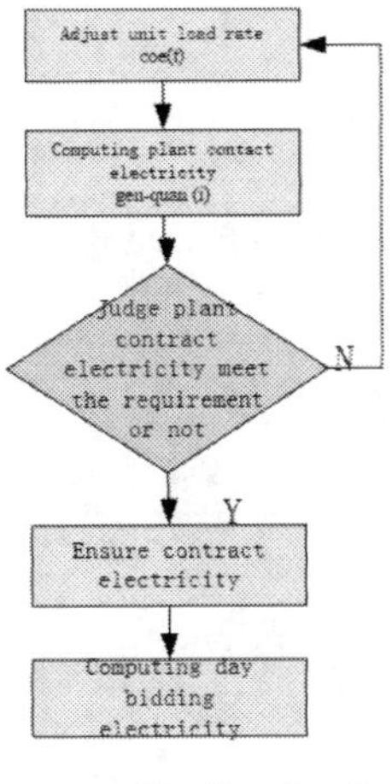

Fig. 2. Contract power allocation algorithm.

According to the existing electricity market operating conditions, the contract power generally accounts for 70% to 90%of the total pre load. The distribution of contract electricity in province is used hourly rolling distribution algorithm based on the daily load forecast, and the contract power is allocated to the corresponding power plants in an hour basis.

The flow chart of the contract allocation algorithm is shown in figure 2. First of all, it can adjust the unit load rate according to the unit actual situation in each power plant, calculate the plant distributed contract power according to load rate, if the contract distributed electric of each plant can meet the demand of actual contract, the calculation should be concluded, whether or not, then it return to adjust the unit load rate. Repeat the above-mentioned process unit the contract power meets the contract requirements.

The calculation steps are as follows:

- Calculated the load rate of plant completed contract electricity at t period time (in hours),

$$\mathrm{coe}(t) = \frac{\mathrm{load}(t) \times \mathrm{coe_{all}}}{\mathrm{gen_{max}}(t)} \qquad (1)$$

In the upper model, $coe(t)$ denote the plant completed contract electricity load rate during the t period, $load(t)$ indicated the system load of all the power plants during the t period, coe_{all} indicated the adjustment factor, $gen_{all}(t)$ indicated the maximum output of all the power plants during t period.

- Determine the power plant electricity contract at contract day:

$$\mathrm{gen\_quan}(i) = \sum_{t=1}^{T} \mathrm{gen\_unit}(i,t) \times \Delta t \qquad (2)$$

In the upper model, $gen\_quan(t)$ is the daily contract electricity quantity of power plant i, $gen\_unit(i,t)$ indicated the basic output of power plant i in time t, Δt indicated time period. It is to determine whether the contract power of all power plants can meet the contract requirements, if the power plant fails to meet the trading contract requirements, then it can adjusted coe_{all}, recalculated (1) until all the power plants meet the requirements.

Calculated $raval_{day}$ the daily bidding power, that is, the difference between the pre load $load_{day}$ and the power plant contract power.

4 Conclusion

The State Grid is implementing power grid reform, it hopes to establish a competitive electricity market, but the present trade is still remained at the provincial level of internal market, it is weak in connections between regions, the trade between provinces electric resource trade is limited.

The electricity market trade model proposed here can support the multi trade between the power generation and the purchasing power within the province and inter province range. In order to maintain power system stability, the trade will be divided into two parts of contract and bidding, the provinces and plant can be considered as two-stage unit, the provinces market complete its inter electricity trade in priority, the surplus-deficit part is balanced by large scale region market. It integrated the provincial and regional trading of electricity markets, and promoted the coordination and distribution of power resources in a large scale, it plays a role of market regulation and promotes the healthy development of the power industry.

Acknowledgment

The authors wish to thank cooperators and other co-workers. This work was supported in part by a grant from the program of "Research and application of key technologies in inter-provincial surplus renewable energy spot market".

References

1. The history of China LJ, Korea electric power market status and prospects of. automation of electric power, **2**, 1-4 (2000)
2. China coal network. New electricity reform over the past two years, the challenges are still many, [EB/OL]., http://www.cctd.com.cn/show-, 19-166296-1.html, (2017)
3. Chen Zheng. Reflections on the construction of regional electricity market. China Energy News, (2017)
4. Tai Xue, Sun Hongbin, Guo Qinglai. Power chain trading and congestion management method based on block chain in power Internet. power system technology, **40** (2016)
5. Zhang Xin Hua, ye Ze. Equilibrium analysis of electricity market. (China Electric Power Press, Beijing, 2013)
6. Xiang Yang. Logic of the structural change of China's electricity market. productivity study, 6, 22-24 (2011)
7. Zou Jun, Zhang Haining, Tang Yi, Li Lei. The blockchain technology guide. (Mechanical Industry Press, Beijing, 2017)
8. Zhu Jizhong. Analysis of the development and implementation of American power market. South China power grid technology, 5, 22-28+101 (2016)
9. Liu Gang. The inspiration of the electricity market reform in Ontario, Canada. China prices, 7, 47-50 (2006)
10. Yuan Jian. Comparison and reference of foreign power market structure. Shandong University, (2014)
11. Cold yuan, Chen Zheng, Ou Peng, Wang Qinghong, Mongolian Sichuan. The latest UK electricity market reform act and Its Enlightenment to China. China energy, 4, 12-15 (2014)
12. Bao Minglei, Ding Yi, Shao Changzheng, Song Yonghua. Review of the Nordic electricity market and its experience for our country. *Proceedings of the Chinese Academy of Electrical Engineering*: 1-11.
13. Luan Fengkui, Jia Junguo, Han Yinghao, Zhou Wenyu, Zeng Zeng. Japan's power industry reform and its reference to China's power market construction. East China power, 9, 6-9 (2006)
14. Weican. Sun Kang Jianing electric power industry reform, the market mechanism and international comparison of. Chinese energy, **25**, 21 (2016)
15. Chang Dongling, Tu Jin, Cheng Xiaolei. Lessons from the American electricity market in California and its implications for the reform of China's electricity market. power technology and economics, 4, 25-27 (2006)
16. CPC Central Committee and the State Council. Some opinions on further deepening the reform of power system. wind energy industry, 4, 7-11, (2015)
17. Luo Juan. Power system reform, No. 9, depth of interpretation, [EB/OL]., http://www.qianzhan.com/analys, t/detail/220/161113-f6400890.html, 2016-11-13.
18. Sales of electricity industry alliance. National electricity reform case of large - sort - Guangdong power market analysis, [EB/OL]., http://shoudian.bjx.com.cn/news/20170527/827867.shtml, 2017-05-27.
19. Sales of electricity industry alliance. National electricity reform case of large - sort - Shandong power market analysis, [EB/OL]., http://shoudian.bjx.com.cn/news/20170526/827629.shtml, 2017-05-26.
20. Sales of electricity industry alliance. National electricity reform case of large - sort - Shanxi power market analysis, [EB/OL]., http://shoudian.bjx.com.cn/news/20170525/827443.shtml, 2017-05-25.
21. Sales of electricity industry alliance. National electricity reform case of large - sort - Anhui power market analysis, [EB/OL]., http://shoudian.bjx.com.cn/news/20170524/827210.shtml, 2017-05-24.
22. Sales of electricity industry alliance. National electricity reform case of large - sort - Sichuan power market analysis, [EB/OL]., http://shoudian.bjx.com.cn/news/20170523/826957.shtml, 2017-05-23.
23. Sales of electricity industry alliance. National electricity reform case of large - sort - Jiangxi power market analysis, [EB/OL]., http://shoudian.bjx.com.cn/news/20170522/826681.shtml, 2017-05-22.
24. National Power Grid Corp. 2017 1-3, power industry operation profile [EB/OL], http://www.sgcc.c, om.cn/xwzx/nyzx/2017/04/339333.shtml, 2017-04-17.
25. Song, song, Current situation and construction of power market in China. electric society, 8, 20-21, (2016)
26. Nakamoto S. Bitcoin: A Peer-to-Peer Electronic Cash System[EB/OL]. https://bit coin.org/bitcoin.pdf, 2008-10.
27. Nakamoto, S., Bitcoin:, A, Peer-to-Peer, Electronic, Cash, System[EB/OL]., https://bit, coin.org/bitcoin.pdf, 2008-10.
28. Narayanan A., Bonneau J., Felten E. Bitcoin and Cryptocurrency Technologies: A Comprehensive Introduction. (Princeton University Press, United States, 2016)
29. Narayanan, A., Bonneau, J., Felten, E., Bitcoin, and, Cryptocurrency, Technologies:, A, Comprehensive,,

https://doi.org/10.1051/e3sconf/20185703001

Introduction, Princeton, University, Press, United, States)

30. Yuan Yong, Wang Feiyue. Block chain development present situation and prospects of. Journal of automation, 4, 481-494, (2016)

Research on the Construction of Security Constrained Economic Dispatch Scenarios for Large-scale Power Grid

Guodong HUANG[1], *Dan* XU[1], *Zhiqiang* LUO[2], *Qiang* DING[1], *Min* LI[3], and *Hongtu* ZHANG[3]

[1]Power Automation Department, China Electric Power Research Institute, Beijing, China
[2]Power Dispatch and Control Center, the State Grid Corporation of China, Beijing, China
[3]Power Dispatch and Control Center, the State Grid SICHUAN Electric Power Company, State Grid Corporation of China, Chengdu, China

Abstract. Large-scale power grid security constrained economic dispatch is an effective technical means to solve the optimal allocation of resources in a large area. Based on the characteristics of dispatching plan in the State Grid Corporation, a processing strategy for optimizing objectives, planning adjustments, and basic data integration is proposed. A method of constructing a large-scale power grid security constrained economic dispatching scenario based on strategies combination is established, and the adaptability of different scenarios and the difficulty in constructing the corresponding model are analyzed. Multi-scenario construction aims at solving different problems in different scenarios and is a useful method for improving the availability and practicality of large-scale power grid security constrained economic dispatching technologies.

1 Introduction

With the rapid development of China's power industry, on the power supply side, large thermal power plants, large hydropower stations, and large wind farms have been gradually completed. The demand of trans-provincial and trans-regional energy consumption has further increased. On the grid side, The construction of UHV interconnected power grids continues to deepen, and the scale of power grids develops rapidly. The structure of power grids has become increasingly complex and the nationwide networking has basically taken shape. The mutual influence and interaction of the power grids at all levels are further enhanced, and the characteristics of the power grid shift from the region mode to the overall mode.

The dispatching plan is an important task concerning the safe and economical operation of the power system and is one of the core businesses of the power control center in China. The supporting technologies in this field include the commonly used SCED (Secure Constrained Economic Dispatch) technology that uses the provincial area as the balance unit. However, this technology has shown some limitations in the trans-provincial and trans-regional energy consumption and multi-level integrated operation of the power grid. One way to solve the above problem is to change the scope of use of the SCED technology and extend it to to a larger grid. However, the large-scale SCED problems will bring great difficulties to data collection, model construction and solution [1-2]. In addition, the objective factors such as the multi-level organizational model of the dispatching center and differences in the balance strategy of the dispatch region will bring more difficulties to practical application.

At present, most of the related researches are started from decomposition and coordination. In [1], a method of decomposition and coordination for large grids based on price response functions is proposed. It decomposes the security constrained economic dispatch model for super-large grids into the main problems of the upper-level tie-lines and the provincial-level unit planning sub-problems. Examples show that this method can effectively improve the economic dispatching efficiency of large-scale power grids, but this method does not consider China's power grids. The actual operation is not strong. Literature [2] proposed a large-scale decentralized unit combination model. In Europe and the United States, due to the continuous expansion of the electricity market, the main problems it faces are the joint optimization between different international and regional market rules, and the focus is on the solution method of the decentralized model. In [3], from the point of view of overall grid coordination, the development of tie line planning seeks to solve the problem of optimal allocation of global resources. However, the simulation example still stays between the two interconnected provinces, and it is difficult to adapt to the large-scale grid problem such as " East China- South China- Central China" interconnection. The literature [4-7] focuses on the study of the solution strategy for large-scale unit combination problems, but does not consider all practical constraints and grid control modes. The literature [8-11] focuses on the study of the absorption of uncontrollable renewable energy sources from models and algorithms. On the whole, the current economic constraints of security constraints on large-scale power grids have not been properly resolved.

According to China's power grid dispatch organization mode, it is currently divided into national-regional and provincial dispatches. National-regional dispatch is mainly responsible for compiling the regional and inter-provincial tie-line plans, which embodies the inter-provincial electricity trading or trans-provincial and trans-regional consumption of large power sources. Provincial dispatch then compile their own unit plans based on load forecasting and exchange plans. According to the actual situation, if we want to carry out the entire grid security constrained economic dispatch, at least we need to consider the following additional issues: (1) optimization target design; (2) large-scale power system balance strategy design; (3) inter-provincial and large-area tie-line plan adjustment principles; (4) processing methods for power supply of different schedule relationship; (5) large-scale power grid network simplified methods applied to the planning requirements. With the deployment of smart grid dispatching and control systems within the State Grid Corporation of China, the basic data collection of dispatching plans in the provinces has been basically solved. This paper mainly proposes solutions to other problems and constructs usage scenarios for large-scale power grid security constraints and economic dispatch in a demand-combined manner. The paper analyzes the data requirements of different scenarios, the problems and the ease of model construction for the "Three Chinas" actual grid model. Enhance the availability and practicability of large-scale power grid security constrained economic dispatch technology under different dispatching modes by means of multi-scenario construction.

2 Strategic design

2.1 Optimize objective design

Usually, the objective of SCED optimization is to purchase the best price for the entire grid and optimize the coal consumption. Taking into account China's actual scheduling strategy, the SCED optimization goals in the province are mostly to consider energy-saving scheduling model of open and impartial power dispatch. However, since SCED users in the entire grid are usually in the overall scope of the decision-making problem, in addition to the provincial "open and impartial power dispatch" optimization goals, according to actual research, SCED of the whole network should add two optimization goals, namely, the regeneration of clean energy consumption and the maximization of power flow margins of key transmission sections.

The optimize objective of open and impartial power dispatch :

$$f_1 = \sum_{t=1}^{N_T}\sum_{i=1}^{N_G}(B_{0,i} + \sum_{k=1}^{K}\int b_{i,k}(p_{i,t})) \tag{1}$$

Where N_G is the number of generator units; N_T is the period of the study cycle, $B_{0,i}$ is the power generation cost constant for the i-th unit, $b_{i,k}(p_{i,k})$ is the power generation micro rate curve for the k-th section of the i-th unit, and similar power generation micro-increasing rates for the same installed capacity, k is the segment number of the multi-stage fee growth rate curve, $p_{i,t}$ is the unit output.

The objective function of maximizing the consumption of renewable energy:

$$f_2 = \alpha\sum_{t=1}^{T}(P'_{W,t} - P_{W,t}) \tag{2}$$

Where α is the penalty coefficient, $P_{W,t}$ is the total amount of wind power actually called by the wind farm during t period, $P'_{W,t}$ is the predicted wind power of the wind farm during period t.

The objective function of maximizing power flow margins of key transmission sections:

$$f_3 = \beta\sum(L_t - L_{t,\mathrm{lim}it}) \tag{3}$$

Where β is the penalty coefficient, L_t is the flow of the section and $L_{t,\mathrm{lim}it}$ is the section limit during period t.

Therefore, there are at least three kinds of optimization objectives that can be selected: (1)Consider the energy-saving of open and impartial power dispatch for thermal power. (2) Maximize reproducible clean energy consumption in the entire grid. (3) Maximize the power flow margin of key transmission sections.

2.2 System balancing strategy

In the traditional province-based SCED model, the balancing strategy is that the total output of the units in the area is equal to the total load demand in the area, as shown in the following formula.

$$P_{load,c,t} = \sum_{i\in c} P_{i,t} - T_{c,t} \tag{4}$$

In the formula, $P_{load,c,t}$ is the total load demand of provincial c at time t; $P_{i,t}$ is the planned output of unit i in province c at time t; $T_{c,t}$ is the outgoing power of provincial c at time t (negative when received) power . Usually $T_{c,t}$ is the data dispatched by the upper-level dispatcher and is balanced according to the determined value.

The above $T_{c,t}$ is a determined value in the model, which is a key factor that can deal with the whole grid decomposition. When the global optimization calculation is performed in the entire grid, it is in fact the optimization of the outgoing (received) power $T_{c,t}$ for each province. A simpler processing method is still to use the balancing strategy of formula (4) to extend the province c to the global grid scope that needs to be optimized. The total provincial load forecast is added as the balance, and the power transmission between the grid and the external grid still uses known values. At this point, when the optimization is completed, $T_{c,t}$ can be obtained through load flow calculation. This approach is simple and straightforward, but the value of $T_{c,t}$ may deviate far from the actual tie-line plan. The reason is that according to the traditional SCED optimization goal, when the power trading related constraints are not added, the tie-line power will be completely determined by the power structure of each province. This is inconsistent with the formation mechanism of China's tie-line plan, which will cause difficulties in its use.

One solution that can solve the above problem is to directly add the constraint of the cross section of the tie-line, and make the tie-line plan within the acceptable range in the constraint form, as shown in the following formula.

$$T_{c,t,\min} \le T_{c,t} \le T_{c,t,\max} \tag{5}$$

Where $T_{c,t,\min}$ and $T_{c,t,\max}$ are the minimum and maximum powers that can be sent (received) at the time t in the province c.

In another solution, the provinces retain the balance constraint and use the tie-line as the equivalent power source to preserve the global optimality. Equivalence means that the physical lines with both ends in the same province are merged into a pair of equivalent power supplies. At this time also meet the following constraints: (1) two-terminal coupling constraints, that is, under the premise of ignoring grid losses, the sum of the power transmitted at both ends of the line is 0; (2) power dispersion injection characteristics, in order to carry out the power flow calculation, the optimize power are distributed at the physical line access points according to the equivalent historical statistics. According to this idea, the SCED scenario of the whole grid needs to add positive-negative coupling equivalent unit information and the power distribution factor.

There are at least three possible system balancing strategies that can be adopted: (1) The unified balance of the whole grid obtained from the traditional provincial balance promotion. (2) Add the entire grid balance of the tie-line constraint. (3) Tie-line participates in system balance as an equalizer.

2.3 Adjustment principle of tie-line plan

In the balance strategy mentioned above, when the tie-line is used as an equalizer to participate in the system balance, the tie-line power is classified as a demand variable, and the emphasis is on the adjustment zone of the given tie-line power. However, in fact, the tie-line plan is mainly based on the cross-regional and inter-provincial balance of large-scale power supply plans and allocation strategies and power exchange between provinces [3]. This shows that the adjustment zone defined for tie-line must meet certain constraints.

According to the actual situation, the power composition of the tie-line can be divided into two parts: provincial power trading and cross-provincial cross-regional balanced power supply, as shown in the following formula.

$$T_{l,t} = T_{l,s,t} + T_{l,u,t} \tag{6}$$

$T_{l,t}$ is the tie-line power; $T_{l,s,t}$ is the power trading decision quantity; $T_{l,u,t}$ is the cross-provincial cross-regional power decision quantity. If the transaction is adjustable, an adjustable interval of $T_{l,s,t}$ can be given based on the rigidity of the transaction. If the large-scale cross-provincial power supply in the entire grid plans to do a fixed output processing (in fact, most of the large-scale cross-provincial power supplies are large-scale thermal power and hydropower groups, and usually plan values are prepared in advance), then $T_{l,u,t}$ does not have adjustment zone. If this part of the power supply is also involved in the optimization, the adjustment interval should be associated with the amount of change in the value, as shown in the following equation.

$$\Delta T_{l,u,t} = \sum \alpha_{i,l,t} \cdot (P_i - p_{i,ini}) \tag{7}$$

$\Delta T_{l,u,t}$ is the adjustment zone of cross-provincial cross-regional power supply decision volume, P_i is plan of the cross-provincial cross-regional power supply i, $p_{i,ini}$ is initial plan of the power supply i, $\alpha_{i,l,t}$ is the power distribution coefficient assigned to the line l for the unit i at time t.

In practice, the tie-line is also divided into large-region tie-line and inter-provincial tie-line. There is also an associated relationship between them. Usually, the power of the line transmission in the region is mostly consumed by multiple provinces within the region. Therefore, the adjustment of the power of the tie-line in the region will inevitably lead to the adjustment of the power of the inter-provincial tie-line in the region. In practice, the power of the regional tie-lines will be apportioned among the intra-provincial tie-lines within the region, so the adjustment relationship is as follows.

$$\Delta T_{a,t} = \sum \gamma_{a,l,t} \cdot (T_{l,t} - T_{l,t,ini}) \tag{8}$$

$\Delta T_{a,t}$ is the adjustment amount of the region tie-line a; $T_{l,t}$ is the plan of the tie-line l in the province at time t; $T_{l,t,ini}$ is the initial value of the tie-line l in the province at time t; $\gamma_{a,l,t}$ is the apportionment factor of tie-line l in the province at time t.

The adjustment method of the tie-line plan is determined by three factors. This paper selects three adjustment models: (1)The power trade can be adjusted, and cross-provincial large-scale power supply plans cannot be adjusted. (2)Both are adjustable. (3) Both are adjustable and take into consideration the relevance of the tie-line between the major regions and the provinces.

3 Basic data integration

At present, the basic data of the whole grid dispatch plan comes from the provincial data reporting, and has now been applied to the quantitative security check of the " East China- South China- Central China", and the verified accuracy of planned flow has been recognized [12]. This has ensured the feasibility of basic data collection to a certain extent. However, due to the substantial differences between SCED and security check, it is not possible to directly copy the way of handling the security check. It is also necessary to consider the solution that is appropriate to the scheduling plan.

3.1 Unit data process

Through the information reported by the provinces, the basic data of the unit is obtained, and the unit data must be specially processed, because the unit has the following special attributes in the dispatching plan: in terms of the scheduling relationship, the unit has local balance unit and cross-provinces balance unit, and different scheduling attributes will determine the way in which it participates in system balancing; in terms of network model construction, the unit has modeling units and unmodeled units, which will determine its impact on network flows and model accuracy. Therefore, in the SCED scenario construction of the entire network, differences need to be considered for the above scheduling attributes of the unit.

For the cross-provincial cross-regional balanced unit, the method of binding with each related tie line is used, as shown in the following formula.

$$T_{l,i,t} = \alpha_{i,l,t} \cdot P_i \tag{9}$$

In the formula, $T_{l,i,t}$ represents the power value of the tie- line l determined by the unit i at time t; P_i is the planned output of the unit i; $\alpha_{i,l,t}$ is the power distribution coefficient of the unit i assigned to the tie-line l at time t. Another issue that needs special attention here is that when the unit power is reflected in the tie-line plan, it can no longer participate in the system balance in the unit form, otherwise it will cause repeat calculation of power supply output. However, it must participate in system power flow calculations, otherwise it will cause power imbalance in the power flow calculation.

For the unmodeled units, the power plant reported as a fixed output processing in the current provincial dispatch plan. In the whole-grid SCED, it also exists in the fixed output mode. Since it does not affect the optimization of the entire grid resources and the power flow analysis, the unmodeled units will adopt the intra-state packaging processing mode, as shown in the following formula.

$$P_{s,c} = \sum_{i \in c} p_{i,s} \tag{10}$$

In the formula, $p_{i,s}$ is output of the unmodeled unit i, this value has been obtained in advance; $P_{s,c}$ is the total output of unmodeled small units within the provincial c. At this point, the total output in the scene replaces the unmodeled small units.

From the analysis, we can see that the processing of the different scheduling relationship units will depend on the choice of balancing strategy: (1) Adopt the unified balance mode of the whole grid, you can not consider the characteristics of cross-provincial cross-regional balance, that is, the unit does not need to be converted into a tie-line plan to directly participate in the balance. (2) Considering the constraint of inter-area and inter-provincial tie-lines, and adjusting according to actual conditions, the different scheduling attributes of the unit must be considered.

3.2 DC transmission system process

In the interconnection of large-scale power grid, the interconnection between regions through a HVDC transmission system is a common approach. Because with the further development of high-power power electronic fully-controlled switching device technology and the research of new control strategies, DC transmission can not only achieve energy transmission, but also make it have certain schedulability characteristics, which is different from the AC connection line. Therefore, it is necessary to handle the HVDC transmission system separately.

The HVDC transmission system is equivalent at the AC converter valve. If the DC power is set as the SCED optimization variable of the whole network, the equivalent unit processing mode is still used. However, compared to the AC line, the DC line plan will be subject to a series of constraints such as DC operating characteristics, methods, and operating procedures. If it is to be included in the optimization, the DC operating characteristics, methods, and procedures must be modeled in detail, converted into unit output caps, and the characteristics of a series of scheduling plans such as the prohibition of operating areas and climbing characteristics should be incorporated into the model constraints. This will be a very complicated issue. We will only give special attention to this issue and will not discuss it in depth.

Therefore, DC should also have the following two processing methods: (1) If the DC system does not participate in the optimization decision, the processing method can be simplified, and processing can be performed according to the fixed output equivalent, but at this time, the optimization space between the areas is reduced. (2) DC takes part in the optimization as a variable, and it is necessary to consider various constraints and it is more complicated.

3.3 Power network model process

For large-scale power grid such as "East China- South China- Central China ", the consideration of the power network model in SCED will bring about enormous calculation [1]. However, if the network constraints are not considered, some of the value of SCED optimization will be lost.

Decomposing large-scale power network into provincial-scale power netwrok using equal-value mode in planning flows is the currently method adopted by single province SCED. If the balancing strategy of the second scenario in Section 1 of this paper is followed, the network can be decomposed into provinces. This will greatly reduce the network analysis calculation time.

On the other hand, since the users of the entire network SCED are mostly dispatch planners, according to the focus of the users, the key lines can be selected to analyze in planning flow. These key lines are mainly distributed in the large power network near area and the near area of the UHV AC/DC, so the lines that need attention are in fact relatively fixed.

First select the specific line according to the principle. Principle One: Pay attention to heavy load lines. The operation of the power grid has a certain degree of continuity. When the mode of operation does not change much, the heavy load condition of the line flow can be determined based on the historical trend and included in the concerned lines. In principle two, when there are major changes in the operation of the power grid, planners can often determine in advance the changes in the flow of critical lines brought about by the changes, so as to determine the lines that need attention. After selecting the specific line that needs attention, you can calculate the branch current according to DC sensitivity analysis, as shown in the following formula.

$$P_{z,t} = \sum \beta_{i,z} \cdot P_i + P_{load,z,t} \tag{11}$$

In the formula, $P_{z,t}$ is the power flow of line z at time t; $\beta_{i,z}$ is the sensitivity of unit i to line z; $P_{load,z,t}$ is the contribution of all loads to this line. Since $\beta_{i,z}$ and $P_{load,z,t}$ are only related to the power network, which can be calculated before SCED optimization.

In dealing with the network model, there are at least the following types of processing: (1) Without considering network constraints, ie without planning power flow analysis, SCED problems are converted into ED problems; (2) Consider the whole network model; (3) Consider the province-wide model according to the whole network; (4) Consider the simplified model of the whole network divided by province.

4 SCED scenario construction for large power grid

Based on the above analysis, large-scale SCED need to face many strategies and basic data processing problems. Different strategies and processing methods will constitute different planning scenarios. Combined with the actual situation and the relevance, this paper has built several typical scenarios for different analysis purpose. The combination method is shown in Table 1.

Table 1. List of scenarios.

Scenarios	**Main Considerations**				
	Balance strategy	***Tie-line adjustment***	***Unit scheduling relationship***	***DC***	***Power network***
1	(1)	(1)	(1)	(1)	(2)
2	(2)	(2)	(1)	(1)	(2)
3	(3)	(2)	(2)	(1)	(2)
4	(3)	(2)	(2)	(2)	(2)
5	(3)	(3)	(2)	(2)	(3)

Through analysis, the following conclusions can be drawn:

Scenario 1 directly extends the traditional provincial SCED problem to the whole network, and is suitable for analyzing the theoretical optimization results obtained in the whole network in accordance with the unit characteristics.

Scenario 2 implements the whole network SCED on the premise of satisfying the power trade adjustment. Because DC is not adjustable, the scenario can be defined as the optimal configuration of resources in the region under the precondition of adjusting the power trade.

Scenario 3 is the optimization of resources in the region under the premise of allowing both power trade and large power optimization adjustment.

Scenario 4 is based on scenario 3, since HVDC transmission is adjustable,it is a resource-wide optimization configuration within the whole network.

Scenario 5 is a resource-optimized configuration within the entire network that best fits the actual situation of the current dispatch in China.

The above lists the various scenarios that this article considers worth studying. It is not a collection of all scenes. Researchers can make different combinations of scenarios based on research needs. Because different scenarios have large differences in the constraints considered by the planning model, building a scenario according to requirements has certain practical significance.

Acknowledgment

Thanks for support from the science and technology program of State Grid Corporation of China (Research on Parallel Coordination Strategy and Key Techniques in Generation Scheduling over Large-scale Power Grids).

References

1. LAI Xiao-wen, ZHONG Hai-wang, YANG Jun-feng et al. Decomposed Optimization Method over Large-scale Power System Based on Price Response Function. Automation of Electric Power Systems, **37** (2013)
2. Mohammad Javad Feizollahi, Large-Scale Decentralized Unit Commitment. August 27, 2014
3. Xu Dan, LI Xiao-lei, DING Qiang Optimization of Tie Line Hierarchical Schedule Based on Network-wide Coordination. Automation of Electric Power Systems, **38** (2014)
4. Wang B, Ye X, Xia Q, et al. Security-Constrained Economic Dispatch With AC/DC Interconnection System Based on Benders Decomposition Method. Proceedings of the Csee, **36** (2016)
5. Ding Q, Li K, Zhu Z, et al. Economic Dispatch Optimization Method for Large-scale Power Grid Based on Variable Dimension Reduction. Automation of Electric Power Systems, **41** (2017)
6. Kargarian A, Fu Y, Li Z. Distributed Security-Constrained Unit Commitment for Large-Scale Power Systems. IEEE Transactions on Power Systems, **30** (2015)
7. Zhao W, Liu M. Cutting Plane Consensus Algorithm for Decentralized Dynamic Economic Dispatch in Multi-area Power systems. Proceedings of the Csee, **36** (2016)
8. Ahmadi-Khatir A, Conejo A J, Cherkaoui R. Multi-Area Unit Scheduling and Reserve Allocation Under Wind Power Uncertainty. IEEE Transactions on Power Systems, **29** (2014)
9. Fu Y, Liu M, Li L. Multiobjective Stochastic Economic Dispatch With Variable Wind Generation Using Scenario-Based Decomposition and Asynchronous Block Iteration. IEEE Transactions on Sustainable Energy, **7** (2015)
10. Li Z, Shahidehpour M, Wu W, et al. Decentralized Multiarea Robust Generation Unit and Tie-Line Scheduling Under Wind Power Uncertainty. IEEE Transactions on Sustainable Energy, **6** (2015)
11. Saurabh Tewari, Charles J. Geyer, Ned Mohan ,A Statistical Model for Wind Power Forecast Error and its Application to the Estimation of Penalties in Liberalized Markets, IEEE TRANSACTIONS ON POWER SYSTEMS, **26** (2011)
12. Men D Y, Ji X, Cai Z, et al. Expert Library-Based Data Verification for National-Regional-Provincial Power Grid Integration Security Correction. Advanced Materials Research, **805-806** (2013)

Study on the Key Problems of UHVAC/DC Power Grids with New Generation Large Synchronous Condenser

Xiao Fan , *Zhou* Kunpeng, *Wang* Tao, *Cao* Kan, *Rao* Yuze

Hubei Electric Power Research Institute, SGCC, Wuhan, China

Abstract. In order to cope with the great pressure caused by the gradually exhaustion of fossil energy and environmental protection and climate warming, the development and application of the renewable energy has become an important foundation and development direction in the field of energy. However, due to the impact of energy and resource endowment, China's load center and renewable energy base into the characteristics of long-range reverse distribution, the use of long-distance large-capacity transmission potential is necessary. With the "strong DC system and weak AC system" problem of the company power grid is increasingly prominent, the higher demand of dynamic reactive power support is put forward in the UHV DC power transmission project. Then, the large-capacity dynamic reactive power of new generation synchronous condenser is large-scale built up in the UHV DC/AC power system.

Due to the high requirement of response speed and capacity in the UHV DC power transmission, the structure, dynamic characteristics, excitation control and relay protection and other aspects of the large synchronous condenser are different from generators and traditional synchronous condenser. Based on this, the dynamic reactive power demand of UHV DC power system is analyzed in this paper. Then, the main situation of large synchronous condenser is considered. In addition, the key points of the excitation control system and protection system are also discussed in this paper. There are important theoretical and practical significance for the safe and reliable operation of the UHV DC/AC power system.

1 Introduction

In order to cope with the great pressure caused by the gradually exhaustion of fossil energy and environmental protection and climate warming, the development and application of the renewable energy has become an important foundation and development direction in the field of energy. However, due to the impact of energy and resource endowment, China's load center and renewable energy base into the characteristics of long-range reverse distribution, the use of long-distance large-capacity transmission potential is necessary. However, the conventional DC converter station needs to consume a large amount of reactive power, which is mainly provided by the static reactive power compensation device in the converter station. The reactive power absorbed by the DC converter station can be significantly increased when a large disturbance is occurred in the power system. While the dynamic reactive power compensation capacity of the power grid will be lacked and the risk of voltage instability is greatly increased due to the output of the static reactive power compensation device is reduced. Compared with the static reactive power compensation device, the capacity of the synchronous power converter compensation is large. In addition, the forced excitation of the excitation system can be implemented automatically under grid voltage is drop condition. Therefore, with the "strong DC system and weak AC system" problem of the company power grid is increasingly prominent, it is of great significance to improve the stability of grid voltage and dynamic reactive power support and very important for the safe and stable operation of UHV AC-DC hybrid power system with the construction process of large synchronous condensers in UHV power system is accelerated.

According to the national grid company planning, there are 66 new generation large synchronous condenser will be deployed in the power system. However, due to the high requirement of response speed and capacity in the UHV DC power transmission, the structure, dynamic characteristics, excitation control and relay protection and the aspects of the large synchronous condenser are different from generators and traditional synchronous condenser.

For the voltage stability, the influencing factors, analytical methods, stabilization criteria and coping measures, the voltage stability problem of DC feed power grid, especially multi-DC feed power grid, are systematically expounded in [1]-[4]. In order to improve the static voltage stability of the receiving power grid and to prevent the transient voltage instability, a load optimization control method is proposed in[5-7].For multiple DC are feed in Shanghai power grid and east China power grid and the southern power grid, the impact of large disturbance are analyzed and the measures, such as, static var compensator (SVC), static synchronous compensator (STATCOM) and synchronous condenser are proposed to improve the transient voltage stability of the power system in the literature [8-11].

For the excitation control system, the excitation system has a significant effect on the fast response performance and reactive power output characteristics of large-scale synchronous condenser. At present, the study of excitation control mainly focuses on synchronous generators. Meanwhile, there is no large-scale synchronous camera excitation control system related to the application practice. In addition, although the main part of of the excitation control system in the large-scale synchronous condenser should be the same with the same capacity generator, but its use of "idling" grid-way, as well as in the low excitation and excitation conditions under the special needs, making new changes have taken place in the excitation control system of the large-scale synchronous condenser.

For the relay protection system, the relay protection configuration scheme of the the large-scale synchronous condenser under the static frequency conversion (SFC) startup mode is proposed. This paper mainly introduces the current and voltage analog quantity input to the relay protection device, which is the obvious low frequency characteristic, and its frequency is changing from 0~50Hz. It is pointed out that the low-frequency characteristics should be considered in the protection configuration of the large-scale synchronous condenser. However, a new generation of large condenser new changes not only have taken place in the startup mode, at the same time in the condenser standardization of relay protection design, configuration and setting technical standards lack of corresponding standards. In addition, there are may be a new problem in the excitation-loss protection of the large-scale synchronous condenser. Based on this, the dynamic reactive power demand of UHV DC power system is analyzed in this paper. Then, the main situation of large synchronous condenser is considered. In addition, the key points of the excitation control system and protection system are also discussed in this paper. There are important theoretical and practical significance for the safe and reliable operation of the UHV DC/AC power system.

2 Voltage stability of UHV power grid

2.1 Voltage stability of DC inverter station

2.1.1 The prominent problem of dynamic reactive power shortage in the end of the power grid

DC system rectifier side using fixed power control, inverter side using the set out arc angle control strategy. Under a voltage drop is occurred in the end of AC system condition, the DC voltage will be reduced. In order to ensure the DC power is constant, the DC current will be increased by the rectifier station. An increase in the DC current will result in a reactive gain increase in the end-fed converter. When the voltage of the AC system is reduced, the reactive power of the AC filter will be reduced. The above two points lead to a decrease in the AC voltage after the converter station from the system to absorb the reactive power increases, the stability of voltage is easy to be lose.

2.1.2 DC feed reduces the system's dynamic reactive power reserve

Under normal conditions, the reactive power exchange between the UHV DC converter station and the system is zero. When 8 million kilowatts of rated power are supplied by DC power, the total active is distributed to the end of the 220kV power grid, and 1.2 million reactive power are consumed by 500 kV step-down transformer. Under the same power consumption condition, if 12 sets of 66 million kilowatts generator instead of DC feed in the end of the grid, about 1.9 million kilowatts of reactive power can be provided by the system. In the dynamic process of the system, a large number of dynamic reactive power can be provided to support the system voltage. Compared to conventional generators, DC large-scale feed reduces the system dynamic reactive power reserve.

2.1.3 DC feed-in deteriorate the voltage system characteristic

Under the AC system voltage decreases and DC feed into the end of the grid condition, the conventional generating units will increase the reactive power output and reactive power will be absorb by the DC inverter station from the system. According to the AC bus voltage reduced by 1%, DC inverter station will absorb 50,000 kvars of reactive power from the system, the same size of the conventional power supply to the system at least 300,000 kvars of reactive power will be provided by the system. Conventional power supply and dc inverter station in the system voltage and reduce the reactive voltage regulating characteristics, on the other hand, the voltage regulation system characteristic are deteriorated due to the DC feed-in.

2.1.4 Dc feed-in reduces the resistance of the system reactive power

As the DC large-scale feed to replace the conventional power supply, the short-circuit capacity decreased in the end of the system, resulting in the system anti-reactive impact capacity weakened in the DC multi-feed area is particularly evident. For example, by the end of the power grid, 8 million kw dc into the equivalent of 12 sets of 660000 kw conventional unit, compared with the conventional unit, ac system short circuit current will reduce about 20000 due to the dc feed-in. According to the reactive power impact, about 60 kV voltage fluctuation will increased due to 4 million dc feed in the backend system. It's means that the system voltage is collapsed.

2.1.5 DC fault (commutation failure) to absorb a large number of reactive power

Under single UHV dc commutation failure condition, there are about 5 million system reactive power will be absorbed by the inverter side. If multiple dc commutation failure at the same time, a large number of reactive power will be absorbed from the system, and even cause damage to voltage stability.

2.2 Short circuit capacity is insufficient in the weak-sending system

With the development of the large energy base extended to outlying regions, especially the mass concentration and development of wind power and photovoltaic (pv), dc transmission sending power grid is weak and short circuit capacity shortage problem is prominent. On the one hand, the promotion of DC transmission capacity is depended heavily on the end of the thermal power, and the flexibility of the operation mode of the power grid and the consumption of the clean energy is impact directly. on the other hand, the transmitter system transient voltage will be rised due to the DC commutation failure, and the large-scale wind energy will be tripped under serious condition.

2.3 Voltage stability problem on the load center

Beijing, Tianjin and other regions of thermal power units will gradually shut down, part of the load from the gas supply unit. At the same time, with the further growth of the load, the proportion of external power will continue to increase. Due to the insufficient contradiction in voltage support, especially in the case of gas pipeline failure, will lead to gas units were forced outage, easy to lead to voltage stability, a serious threat to the power supply security in the capital area.

In summary, with the rapid development of UHVDC, large-scale development of clean energy, a large proportion of the concentration of power areas, the company's power grid characteristics of a large change in some areas of dynamic reactive power reserve decline, the problem of insufficient voltage support is prominent. Voltage stability problem has become one of the main problems of large and small power grid security and stability. It's means that the large-scale active DC transmission must match large-scale dynamic reactive power, that is, "large DC transmiss, strong reactive support".

In order to solve this problem, a large number of analytical studies have been carried out in the aspects of DC system voltage stabilization mechanism, AC / DC short circuit ratio, multi-feed transient voltage stabilization, dynamic reactive power support capability, generalized voltage safety warning and control system. The final scheme of the State Grid Corporation is decided to build a large-scale synchronous condenser to solve the voltage stability of the AC-DC hybrid system.

3 Basic situation of the large synchronous condenser

Synchronous condenser (also known as synchronous compensation machine) is a reactive power compensation device, is running in the motor state, to the power system to provide or absorb reactive power synchronous motor, without mechanical load without the prime mover. The grid structure and the secondary system of the large synchronous condenser is shown in figure 1.

Due to the high requirement of response speed and capacity in the UHV DC power transmission, the structure, dynamic characteristics, excitation control and relay protection and other aspects of the large synchronous condenser are different from generators and traditional synchronous condenser.

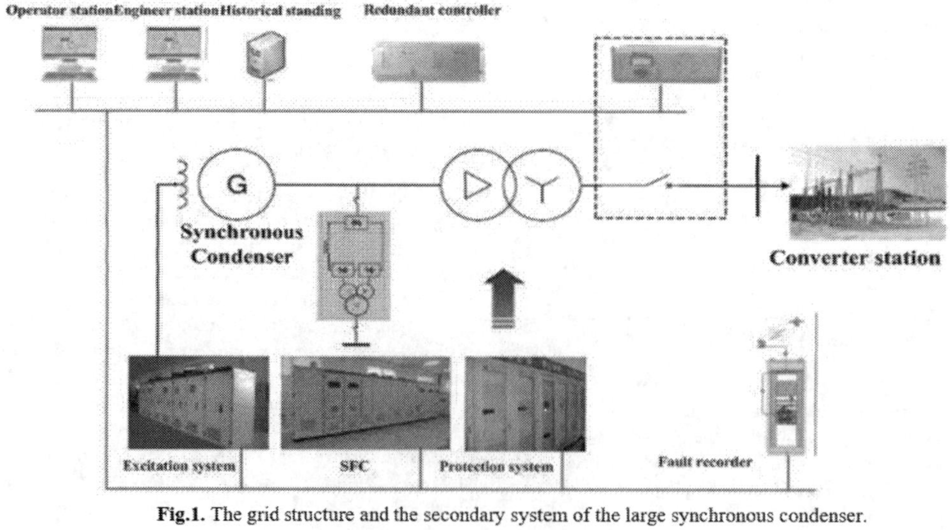

Fig.1. The grid structure and the secondary system of the large synchronous condenser.

https://doi.org/10.1051/e3sconf/20185703003

4 Start mode of the static frequency conversion (SFC)

Start excitation with the SFC, the unit is draged to 1.05 times the rated speed. The SFC exit is operated and the rated terminal voltage is established by the main excitation system. Then, the unit under idling process condition, the same period at the same time point of the grid. Starting mode of the SFC can be seen in fig.2. The process is as follows:

Step (1): under the cooperation of the excitation system, the SFC tractor unit rotor is accelerated by the turning state (except for the fast restart). During the start-up process, the excitation system adopts its excitation circuit. Step (2): If the rotor of the large synchronous condenser is accelerated to 52.5Hz, the SFC system will exits. Then, the rotor began to degenerate due to loss of driving force. Step (3): In the process of unit inert speed down, excitation system will switch to the main excitation (the shunt) circuits and the start excitation system to exit. Step (4): After confirming that the SFC has been reliably disconnected from the stator, the excitation system enters the boosting process and establishes the rated terminal voltage. Step (5) :in the subsequent idle speed process, by the quasi-synchronization device to complete the same point of time to capture, and phase control unit into the grid.

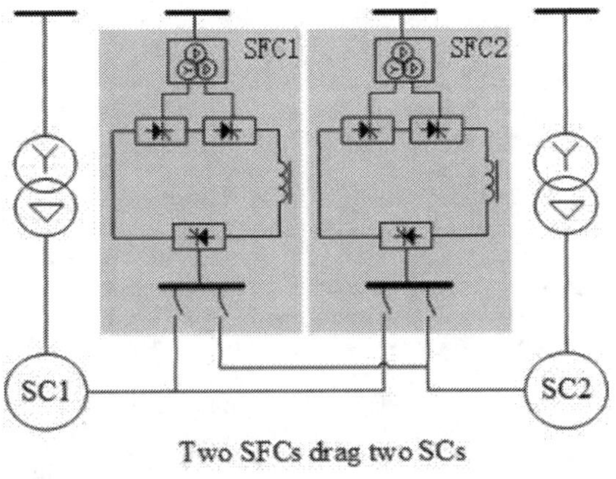

Fig.2. Starting mode of the SFC.

5 Excitation control system

There are differences between the large-capacity tuner excitation system and generator excitation model, especially the extreme operating capacity of the large synchronous condenser under the low excitation and over-excitation conditions are different with the synchronous generator. Therefore, it is necessary to establish the special control model such as low-excitation and over-excitation of the large-capacity tuner excitation system. This paper studies the operation control strategy of the large capacity modulator under abnormal operating conditions and its exit mechanism under the extreme working condition, so as to give full play to the support of the large-scale modulator in the special condition of the grid voltage stability and dynamic reactive power.

5.1 Performance requirements of excitation system

Under power grid fault conditions, 3.5 times excitation voltage need to be provided when the terminal voltage is 0.8p.u.At the same time, the top value of the current needs to meet not less than 2.5 times, the time is not less than 15sec requirements, and its excitation voltage response time needs less than 0.1sec.

5.2 Control mode

Under steady state condition, the steady-state operation is control by the high-voltage bus voltage and reactive power, and the transient operation is control by the voltage closed loop for fast excitation or reduction. Main ring of excitation control system is shown in fig.3.

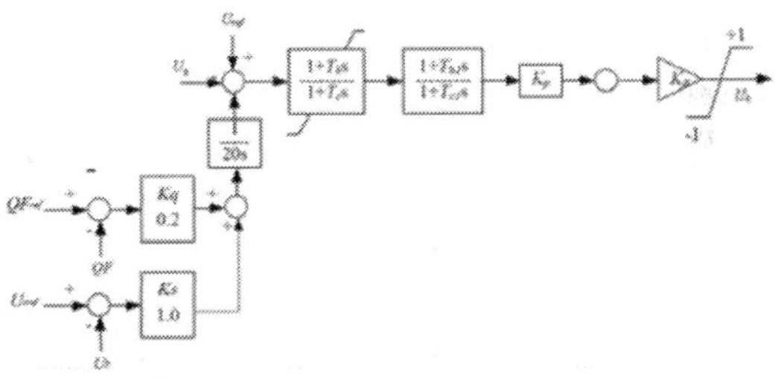

Fig.3. Main ring of excitation control system.

In addition, the fast reactive power control links is increased. The converter filter is cooperate with the synchronous condenser during the switching process. After received for the instruction, the Qref q is the the aim to regulate in the synchronous condenser to achieve rapid reactive power output. The fast reactive power control links of excitation system are shown in fig.4.

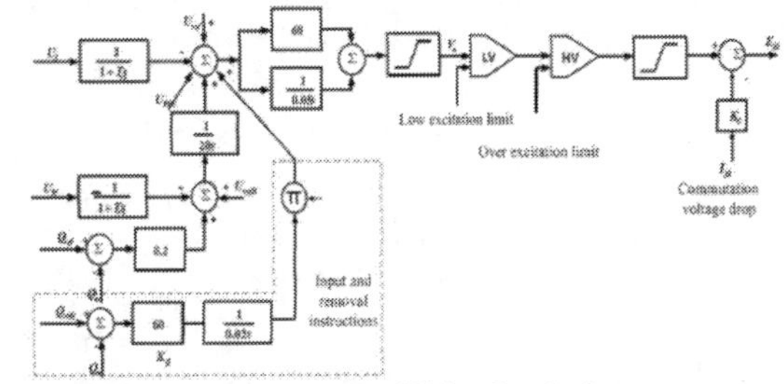

Fig.4. Fast reactive power control links of excitation system.

In addition, for the sending and receiving end grid, the characteristics of the high and low voltage operation under serious fault conditions, the influence of the structure and parameters of the excitation system on the rapid response performance of the large-scale synchronous condenser in different operating conditions

 https://doi.org/10.1051/e3sconf/20185703003

needs to be considered. At the same time, for the many control strategies in the DC system rectification side and inverter side, the excitation control strategy of the synchronous condenser needs to meet the demand of rapid reactive power in the sending and receiving end power grid. The contrast of reactive power in synchronous condenser with or without cutting strategy can be seen in fig.5.

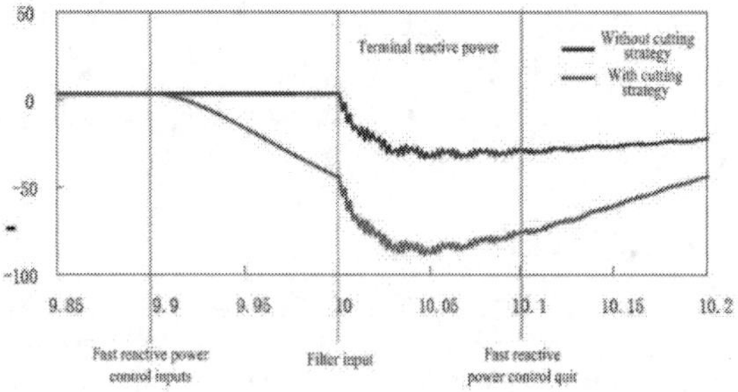

Fig.5. Contrast of reactive power in synchronous condenser with or without cutting strategy.

6 Configuration and principle of relay protection

The protective function of the large synchronous condenser are basically the same with the conventional generators. However, due to the special start-up and operation mode of the large synchronous condenser. The start torque of the SFC is provided by the synchronous condenser. The synchronous condenser is drag by stationary state to the rated speed condition. The corresponding unit electrical frequency varies rapidly from zero Hz to the rated frequency. After the synchronous condenser is connected to the grid, there is phase and late mode of operation, and the reactive power will be absorbed or send. Therefore, the principle and function configuration of the conventional protection group protection cannot be fully applied to the phase adjustment unit, is still need further study and analysis. Especially, in the protection in start process, loss of magnetic protection principle, as well as the protection in power grid and other aspects need further study.

6.1 Protection on SFC start-up process

Due to the large synchronous condenser is started with SFC, the low frequency characteristics of the electricity are present in protection device during start-up process. During the start-up process, the configuration and implementation of the protection scheme on the large synchronous condensers need to take into account the effects of low frequency characteristics. Some special functions can be configured to low frequency condition, such as, the low-frequency over-current protection, the differential protection and the stator ground fault protection in start-up process. Therefore, the independent of the frequency of the signal is necessary to be adopt in protection algorithms. In addition, the other protection of the large synchronous condenser may be mis-operation in the start-up process. Therefore, some measures can be used to prevent the mis-operation of the protection of the large synchronous condenser on start-up process.

6.2 Loss of magnetic protection

Conventional synchronous generator loss of magnetic protection of the stator impedance criterion is the use of static impedance impedance circle or asynchronous garden. However, the synchronous condenser has the ability to run without excitation. As a result, there will be no static stability and out-of-step problems when the synchronous condenser is out of force. In fact, the non-excitation operation of the synchronous condenser will absorb a lot of reactive power from the system. In order to avoid a large number of reactive power converter caused the system power grid decline or even voltage instability problems, the configuration of the loss of magnetic protection can be constituted by the reactive components and the low-voltage components. In addition, in order to prevent the oscillation, short circuit or voltage circuit breakage and other abnormal circumstances of the protection of the situation, the corresponding latching components can be configured for the synchronous condenser.

6.3 Protection of the grid with large synchronous condenser

Due to the large-capacity two-way dynamic reactive power of the synchronous condenser, the original grid voltage distribution characteristics of the power system will be changed with the synchronous condenser. In particular, under the bidirectional dynamic reactive power condition, the inductive current or capacitive current will be provided by the synchronous condenser. It's means that the current distribution characteristics of the grid are changed. It could have new effects on AC current protection, distance protection and voltage protection. At the same time, the operation of the HVDC requires a lot of reactive power, the commutation voltage of the DC system can be maintained due to the Reactive power from the synchronous condenser. In addition, the probability of the inverter commutation failure is reduced when the bus voltage of the AC system is greatly reduced. Therefore, the impact of UHV DC protection with the large synchronous condenser is also an important concern.

In addition, compared with traditional synchronous generator, the standardization of relay protection design, configuration and setting technology of the large synchronous condenser are also changed. The standards for in a timely manner to carry out the work.

7 Conclusion

Due to the high requirement of response speed and capacity in the UHV DC power transmission system, the structure, dynamic characteristics, excitation control and relay protection and other aspects of the large synchronous condenser are different from generators and traditional synchronous condenser. Based on this, the dynamic reactive power demand of UHV DC power system is analyzed in this paper. Then, the main situation of large synchronous condenser is considered. In addition, the key points of the excitation control system and protection system are also discussed. Finally, the simulation model of the new-generation synchronous condensers access to UHV DC power grids are built based on the PSCAD software. The digital simulation results validate the transient reactive power support ability of the large synchronous condenser to UHVDC power transmission system. There are important theoretical and practical significance for the safe and reliable operation of the UHV DC/AC power system.

Acknowledgment

This work was supported by the State Grid Headquarters Science and Technology Project: Research on Fault Analysis, Relay Protection Technology and Operation Control Technology of the large synchronous condenser.

References

1. LIU Zhenya, ZHANG Qiping, WANG Yating, DONG Cun, ZHOU Qinyong. Research on Reactive Compensation Strategies for Improving Stability Level of Sending-end of 750 kV Grid in Northwest China. Proceedings of the CSEE, **35** (2015)
2. Xu Zheng. Dynamic Behavior Analysis of AC / DC Power System. (Mechanical Industry Press, Beijing, 2005)
3. Zhang Jianshe, Zhang Yao, Zhang Zhichao, et al. Influence of dc system control modes on voltage andpower recovery after large disturbance in hybrid ac/dcsystems. Power System Technology, **29** (2005)
4. Yang Weidong, XuZheng, Han Zhenxiang. Coordinatedrecovery strategy for multi-infeed hvdcsystems. Electric Power Automation Equipment, **22** (2002)
5. Yang Weidong, XueYusheng, Jing Yong, et al. Acoordinated recovery strategy for 3 hvdc systems involvedin the south china power grid. Automation of ElectricPower Systems **27** (2003)
6. Wu Ping, Lin Weifang, Sun Huadong, et al. Research and electromechanical transient simulation on mechanism of commutation failure in multi-infeed hvdc power transmission system. Power System Technology, **36** (2012)
7. Ren Jing, Li Xingyuan, JinXiaoming, et al. Simulationstudy on commutation failure caused by switching acfilters of inverter stations in multi-infeed hvdcsystems. Power System Technology, **32** (2008)
8. JinXiaoming, Zhou Baorong, Guan Lin, eta. HVDC-interaction-strength index for the multi-infeed-hvdc power system. Automation of Electric Power Systems, **33** (2009)
9. Li Weifang, Tang Yong, Bo Guangquan. Definition and application of short circuit ratio for multi-infeed ac/dcpower systems. Proceedings of the CSEE, **28** (2008)
10. Sun Jingqiang, GuoXiaojiang, Zhang Jian, et al. Dynamic characteristics of receiving-end of multi-infeed hvdcpower transmission system. Power System Technology, **33**(2009)
11. Zhou Dongxia,YuRui, etc. SFC start and relay protection of 300 Mvar large synchronous condenser. Power System Protection and Control, **20** (2016)

The Efficiency Assessment of Transmission Grid with Massive Clean Energy Based on Production Simulation

Zhao Lang[1], *Li* Jian[2], *Zhang* Pengfei[2], *Peng* Dong[1], *Liu* Lixia[3], *Liu* Shuyong[3], *Long* Wangcheng[1], *Kong* Xiangyu[4]

[1]The center of transmission grid planning, State Power Economic Research Institute, Beijing, China
[2]The department of development planning, State Grid Corporation of China, Beijing, China
[3]The Economic Research Institute, State Grid Tianjin Electric Power Company, Tianjin, China
[4]The department of electrical engineering, Tianjin University, Tianjin, China

Abstract. As the clean energy developing rapidly in recent years, the operational mode of power system was adjusted persistently because it was impacted by the randomness of clean energy output. Meanwhile, the efficiency of transmission grid was decreasing in certain level for the same reason. The production simulation model of power system was presented, and the power source which considering the output characteristics of different clean energy was described respectively. In additionally, the analytical method of efficiency assessment was proposed to assess the impact degree of clean energy. Moreover, the efficiency index of power transmission system programming was adopted to choose planning scheme. At last, three planning schemes of a power grid with massive clean energy are analyzed to confirm the production simulation model and assessing method.

1 Introduction

With the global climate warming and ecological environment deteriorating seriously, the carbon emission problem becomes more and more concerned in the worldwide. Eventually, in order to protect the climate stable relatively, the utilized of fossil fired energy was controlled orderly all around the world. Therefore, several protocols such as the Paris Climate Accord COP21 have gone into effect. There are two main ways to reduce the carbon emission, one is developing clean energy to take place of traditional fossil energy; another is saving energy to cut down the consumption. After years of effort, the percentage of fossil-fired energy decreased for the 21st century due to renewable energy explosive growth [1]. As report by IEA, in 2016, total combustible fuel plants accounted for 60.8% of total OECD (Organization for Economic Co-operation and Development) gross electricity production (made up of 57.8% from fossil-fuel-fired plants and 3.2% from biofuels, waste, etc. plants), nuclear plants 18.0%, hydroelectric plants 12.9%, and geothermal, solar, wind, tide and other plants at 8.2%. The fossil consumption of China decreased around 320 million tons in 2016. The most important factor about demand falling in China was lower consumption of electricity generation, although China still uses half the world's coal. Alongside the decrease of fossil-fired generation, 2016 also saw the continued increase of renewable generation across the OECD and in countries like China. In the OECD, renewables generation grew by 3.8% to account for 23.8% of all electricity generated, its highest share to date [2]. However, the output of hydroelectric energy and wind energy variate disorderly at any time, the power system operation has been more complicated by massive clean energy increment. Moreover, the power system are required to construct or extend timely to satisfy the new clean energy power accessing to grid, the reasonable power grid planning of clean energy integration ought to consider the security, the efficiency and the social benefit. The tradition power grid planning method estimates the schemes based on a maximum operation mode or minimum operation mode, it cannot reflect the randomness output of clean energy and the influence to other devices. Because of those, a simulation method which is capable to apple to the randomness problem is obliged to be present for evaluate the power grid planning.

The stochastic nature of wind generation increases the fluctuation and the uncertainty on the net load. How well simulation the system running was crucial in analysing the planning scheme. H.T. Yang and S.L. Chen [3] presented a multiobjective optimization approach to generation expansion planning. G. A. Orfanos, P. S. Georgilakis and N. D. Hatziargyriou [4] proposed an efficient approach for probabilistic transmission expansion planning that considers load and wind power generation uncertainties. J. Hargreaves, E. K. Hart, R. Jones, A. Olson [5] outlined a new methodology for modeling the economic tradeoffs in implementing flexibility solutions for integrating renewables. M. Lin [6] presented the results of an investigation of the relative computational speeds and solution qualities of six different probabilistic production cost simulation methods for power systems. W. Wu, Z. Hu etc. [7] proposed a new heuristic methodology for Transmission network expansion planning based on chronological evaluation. A.S.D. Braga, J.T. Saraiva [8] presented a multi-criteria formulation for multiyear dynamic transmission expansion planning problems. F. Bouffard, F. D. Galiana [9] formulated a short-term forward electricity market-clearing problem with stochastic security capable of accounting for non dispatchable and variable wind power generation sources. R. Billinton, W. Wangdee [10].made bulk electric system reliability analysis associated with large-scale wind farms, which assists system planners to create potential transmission reinforcement and schemes to facilitate large-scale wind energy conversion systems additions to a bulk system. Tuohy and O'Malley [11] used Wilmar planning tool which incorporates a stochastic unit commitment model considering reserve for wind and load forecast errors. Ummels et al. [12] relied on the chronological simulation program, PoerSym3, which solves an electricity market-based UC-E model using a heuristic algorithm. Black, et al. [13], [14] developed a methodology that incorporates day-ahead, real-time dispatching, and stochastic sampled power imbalances with reserve allocation. Zhang Ning, et al. [15, [16] describes a chronological production simulation platform and shows how it has been applied to the planning of pumped storage. An important characteristic of these studies is that the simulated system operation day-to-day to be able to capturing the system congestion cost as accurately as possible [17]. The previous studies usually aren't really aware of the efficiency of planning schemes, which is particularly significant which reflects the assets efficiency among different undetermined schemes.

The sequential operation simulation model was established to effectively estimate operation condition of future power system. The model considered load distribution, generator outage and generation dispatch algorithm. Transmission congestion in power system was calculated, the effect of alleviating transmission congestion was analysed. Finally , the efficiency evaluation method of power grid based on improvement was proposed to be used in choose planning scheme..

2 Production simulation model

2.1 Simulation framework

The production simulation framework was used to assess the efficiency of power grid contains seven part, and those was shown in Fig.1. In that, the input-data function is completed in Preparing Data module, the main role of power grid model, load model and source optimization model is to prepare the basic data for Load Flow Calculation module. The efficiency index can be obtained from calculation's results. As a result, the efficiency levels are evaluated by a series of methods, and the different planning schemes are assessed in planning evaluation module by efficiency level and economic.

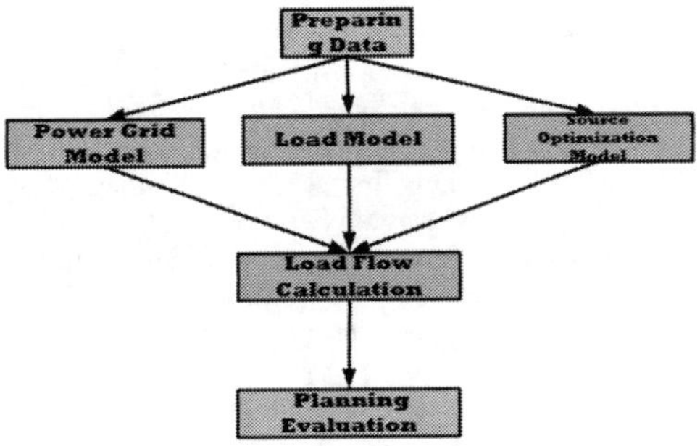

Fig. 1. The framework of production simulation

- Preparing Data

The production simulation requires serval kinds of detail data to describe the power system information. Those might be classified into three types, that is, equipment parameter, operation data and predictive information. In which, the equipment parameter is made up of generations, transforms, transmitting line and the price of each kinds of power; the operation data contains the active power of every lower voltage transformer at each hour in whole year, the hourly wind speed at the wind farm for several years, the hydro energy at every hour of the run-of-river hydroelectric station for several years, the photovoltaic power station output at every hour for several years; the predictive information mainly includes the load growth rate, the installed equipment in planning scheme and the cost of each kinds of power.

- Grid Model

The function of power grid model is constructing the network connection hour-by-hour during the simulation. There are three steps to take shape new connection relation for calculation, the first step is constructing the initial network frame of simulation, the devices have put into operation before simulation date must join to the network frame, the second step updates the network frame hour-by-hour according to the new project's commissioning date during simulation. As often as the new project put into operation, the connection relation adjusts timely based on the system scheme. The above-mentioned new projects cover being constructed and planned. The third step is embodying the maintenance schedule and the operation of power grids, the component during maintenance stage is invalid in network connection, and the equipment arranged as standby application is invalid in network connection. The network frame of every time during simulation is formatted.

- Load model

The fluctuation of the load is expressed in the sum of load from the lower voltage transformers. Therefore, the load model is specific to every lower voltage substation in simulation, and it determines each transformer load characteristic during simulation. The load curve for a transformer might be represented by a sequential array.

For the power grid with N transformers, the load characteristic in whole yare is represented by a matrix P (N, 8760). The load of power grid is summed by each transformer. But to calculated the planning load distribution, those are divided to operated transformer and planned one, and their formulas are disparate owing to the difference in load growth. For operated transformer, the load characteristic in planning year is almost similar as the one in present, that might increasing in some certain rate according to the development of local economic and society. However, for the new operating transformers, those mainly satisfy the new appeared or existed load in local area. Therefore, the transformer load could calculate as Equation (1) and (2), Equation (1) is for operated transformer and Equation (2) is for new operating transformer.

$$P(\mathrm{u},j)= P_{now}(\mathrm{u},\ \mathrm{j})\times(1+v\%)^{(Yp-Ynow-1)}\times n/(n+m(\mathrm{u})+k(\mathrm{u}) \tag{1}$$

In which, $P(u, j)$ is the even load of transformer j at u hour at planning year, $v\%$ is the growth rate of electricity consumption in the area of transformer located on, Y_p is the planning year, Y_{now} is status year, n is the number of transformers in status year, $m(u)$ is the number of operating transformers under contracture at u hour, $k(u)$ is the number of operating transformers in planning scheme at u hour.

$$P_{p1}(x)=\sum P_{now}(\mathrm{u},\ j)/\mathrm{n}\times(1+v\%)^{(Yp-Ynow-1)}\times(m(\mathrm{u})+k(\mathrm{u})/(n+m(\mathrm{u})+k(\mathrm{u}) \tag{2}$$

In which, $P(u, x)$ is the even load of transformer x at u hour at planning year, transformer x is the one put into operation in planning year, $v\%$ is the growth rate of electricity consumption in the area of transformer located on, Y_p is the planning year, Y_{now} is status year, n is the number of transformers in status year, $m(u)$ is the number of operating transformers under contracture at u hour, $k(u)$ is the number of operating transformers at planning scheme at u hour.

Both in Equation (1) and (2), the transformers in equation is lower voltage in power grid. The $m(u)$ and $k(u)$ is related to the simulation times, for example, $m(1000)$ is equal to 3, it represents that there are 3 transformers which is under contracture at present will have put into operation at 1000 hour in simulation year. In other word, $m(u)$ will add 1 until a transformer put into operation at u hours, if there is no transformer start operating at u hour, the $m(u)$ is equal with $m(u\text{-}1)$. As a result, the load matrix of every transformer is gained by a series of calculation.

- Generation model

Generally, the generation output should be keep balance to load levels, and it is required keep necessary reserve for power system stable operation. The generation model considers various constraints of different energy. Those are divided into four categories.

1) Gas turbines, it can be stopped and restarted daily and it on/off state is modeled on an hourly basis.
2) Nuclear plants, large thermal unit, and combined heat and power units, those cannot be cycled daily, therefore, they are modeled as being in on/off state at least an entire day.
3) Wind farms, solar plant and run-of-river hydroelectric station, their capacity is determined by the primary energy which can be obtained through long term statistics, and the output be reduced due to minimum load contains or transmission constraints, the reduction can be recorded to reflect the abandoned condition of clean energy. The scholars [18, 19] presented the PV generation character under planning analysis.
4) Hydro units, it has maximum energy constrains and can adjust output hourly, so it is modeled on an hourly basis.

Equations (3)-(9) describe the constrained; the generation's output vectors for every hour in simulation year are calculated by adopting optimization method.

$$Min\sum_{i=1}^{M}\mathrm{C}\ (\mathrm{x}_{i,t})+\sum_{i=1}^{M}s_{it} \tag{3}$$

$$\mathrm{S.\ T.}\ P_{gi,\min}\le x_{i,\ t}\le P_{gi,max} \tag{4}$$

$$\sum_{i=1}^{M}x_{di}=\sum_{j=0}^{n+k+m}P(\mathrm{u},\ \mathrm{j}) \tag{5}$$

$$0\le x_{i,t}-x_{i-1,t}\le P_{up-i}\ \mathrm{or}\ 0\le x_{i-1,t}-x_{i,t}\le P_{down-i} \tag{6}$$

$$\mathrm{T}_{off}(\mathrm{i})\le \mathrm{t}_{i-off,i} \tag{7}$$

$$\mathrm{T}_{on}(\mathrm{i})\le \mathrm{t}_{i-on,i} \tag{8}$$

$$\mathrm{x}_{i,\ t}=0,\ \mathrm{t}\in(\mathrm{U}_{1i},\ \mathrm{U}_{2i}) \tag{9}$$

$$-S_m\le\sum_{c=1}^{w}GSDF_{c,m}\times x_{i,t}+\sum_{d=1}^{z}LSDF_{d,m}\times P_{b,t}\le S_m \tag{10}$$

The objective is the minimization of the total system operation cost, shown in Equation (3), it mainly includes start-stop and generating fuel cost $C(x_{i,t})$ is the generation fuel cost with the output vector being $x_{i,t}$, the $s_{i,t}$ is the start-stop cost under those circumstance. Equation (4) enforce the limits on output of each generator, $P_{gi,max}$ and $P_{gi,max}$ are the maximum output of the generation and the minimum one. Equation (5) represents the load-generation balance constrain, the x_{di} is the output of generator, $P(u,j)$ is the load of transformer j. Equation (6) enforces the generation ramp rate limits, $P_{up\text{-}i}$ and $P_{down\text{-}i}$ is the positive ramp limit and negative one respectively. Equation (7) and (8) define the states of the generators as discrete variables, $T_{off}(i)$ and $T_{on}(i)$ is the least time of generation i. Equation (9) enforce the maintenance state of generator, (U_{1i},U_{2i}) represents the maintenance time of generation i. Equation (10) represents the restrain of power grid, S is the limit capacity of branch m included transmission line and transformer, $GSDF_{c,m}$ is the generation shifted distribution factor of branch m, and $LSDF_{d,m}$ is the load shifted distribution factor of branch m, $x_{i,t}$ is the output of generation i at t hour, $P_{b,t}$ is the load of substation b at t hour. Ultimately, the units output vector are optimized for calculation.

- Load flow model

The load flow model adopts the DC-flow method based on network frame, load matrix and units output vector being known, the load flow of every device in power grid is gained through DC-load flow calculations.

- Planning scheme evolution

A series of indexes for evaluating contains efficiency and benefit of the proposed planning scheme through power system production simulation. The efficiency index includes the utilization ratio of facility, which is about transformer and transmission line, the heavy load proportion of facility, and the light load proportion of that; the benefit indexes comprises cost of power supply, line loss, and abandon clean energy. The evaluation method for power system planning schemes is improvement comparing the planning scheme and present levels, in other words, the recommend scheme must be the one with the highest improvement.

2.2 Simulation process

The power system production simulation process is built for assessing the efficiency of planning scheme, as illustrated diagrammatically in Fig.2. After setting times and preparing data, the necessary elements to simulation are provided on account of proposed model in section 2.1. Then, the calculation is executed according to simulation times, the simulation adopted the DC-flow method based on network frame, load matrix and units output vector which being known by corresponding model, the efficiency indexes is gained through the load flow of every device in power grid from simulation.

The electric Power System Production Simulation Station (EPSPST 1.0) was developed by ourselves, which was adopted VC programming and matrix computing library.

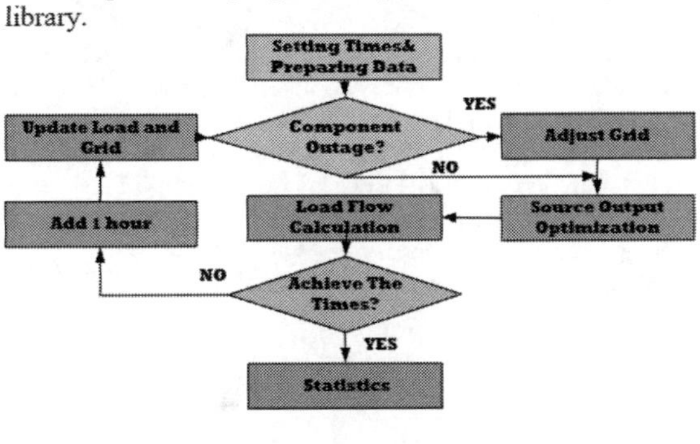

Fig. 2. The process of production simulation

3 Experimental situation

3.1 Abbreviations and acronyms

Tanggu distinct in Tianjing power grid is used as an experimental example to confirm the practicability of production simulation model in this paper. The power grid was made up of two 220kV substations of with 5 transformers and ten 110kV substations with twenty two transformers, six 220kV transmission lines and thirty nine 110kV transmission lines in system until 2017. Meanwhile, two 110kV projects are under construction or in the plan. To 2020, those includes one 220kV substation with two transformers, two 110kV substation with five transformers, three 220kV transmission lines and twelve 110kV transmission lines would put into operation. The raw data is the statistical results from actual running data and simulation data, and the index calculation ways was introduced in our previous study [20].The efficiency index of the present-day power grid is lied out in Table.1.

Table 1. The efficiency index of present-day power grid.

Index	110kV	220kV
Utilization ratio of Tran	25.5%	27.6%
Heavy load proportion of Tran	1.2%	0.45%
Light load proportion of Tran	5.2%	4.2%
Utilization ratio of Line	30.6%	24.3%
Heavy load proportion of Line	1.1%	0.34%
Light load proportion of Line	6.1%	4.5%
Cost of power supply(g/kWh)	305	
Line loss	1.4%	
Abandon clean energy proportion	1.5%	

3.2 Generation

Nowadays, two 220kV thermal power plants contained 300MW coal-fired, 600MW gas-fired and 300 MW hydro plant in power grid, a 50MW wind plant that connect to 110kV power grid through step-up transformer. At the same time, a PV plant is planned to operation at 2020. Moreover, there are two 120MW wind plants and another 60MW PV plant is suggested to construct in planning phase, the wind output and solar irradiation characteristic curves are shown in Fig.3, in which, the typical wind plant output in one day of each month is shown in Fig.3 (a), the average day output of wind plant in a year is shown in Fig.3 (b), and the PV plant output is shown in Fig.3 (c) and (d).

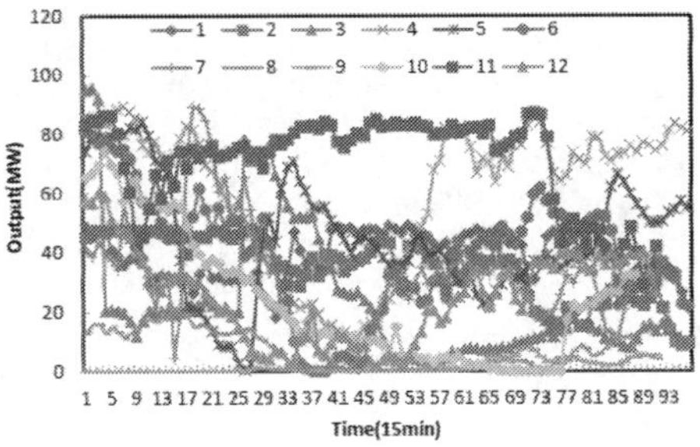

(a)

https://doi.org/10.1051/e3sconf/20185703004

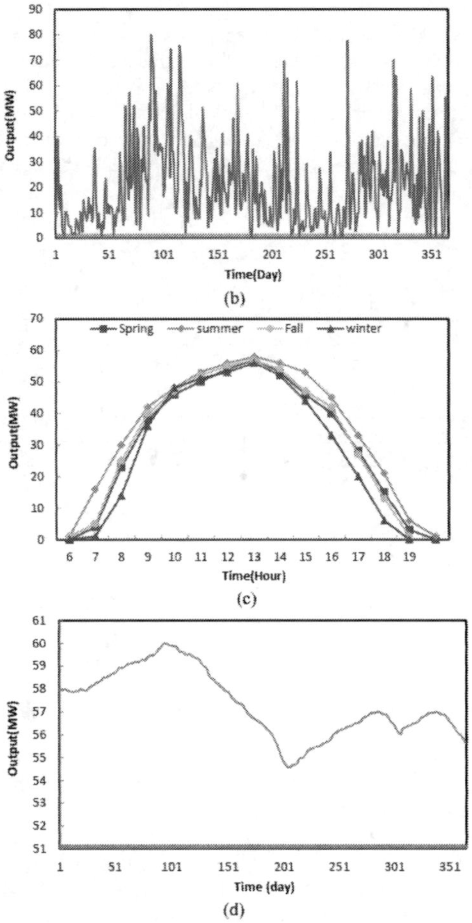

Fig. 3. The typical continuous output of wind plants (a, b) and PV plant (c, d) in the district.

3.3 Load

The total load of 110kV step-down transformers is 1090MW and the electrical power of that were 51 billion kilowatt hours in 2017. The total load and electrical power of every transformer is forecasted according the requirement of that area. The average increment rate of the consumption is about 3% per year. The load curves of the power grid at last year are shown in Fig.4, in which, the average load, maximum load and minimum load of everyday in a year is shown in Fig.4 (a), and the typical day-load curves of each month is shown in Fig.4 (b).

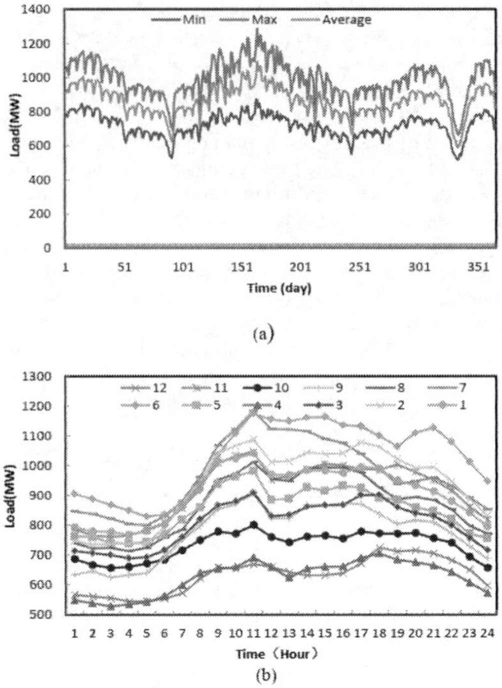

Fig. 4. The load curves of a whole year. (a) and a day (b)

3.4 Planning schemes

In the next two year, totally 300MW clean energy will connect into power grid. For that, three planning schemes to access to the clean energy were proposed, that named as A, B and C. the scheme A was building four 110kV step-up substation respectively and connect into neighboring 110kV substation with each 110 kV transmission line. The scheme B was constructing one 220kV step-up substation and two 110kV step-up substations and access to nearby substation. The scheme C was established two 220kV step-up substations and those transmission lines, the schematic diagram of 3 planning schemes was shown in Fig.5.

https://doi.org/10.1051/e3sconf/20185703004

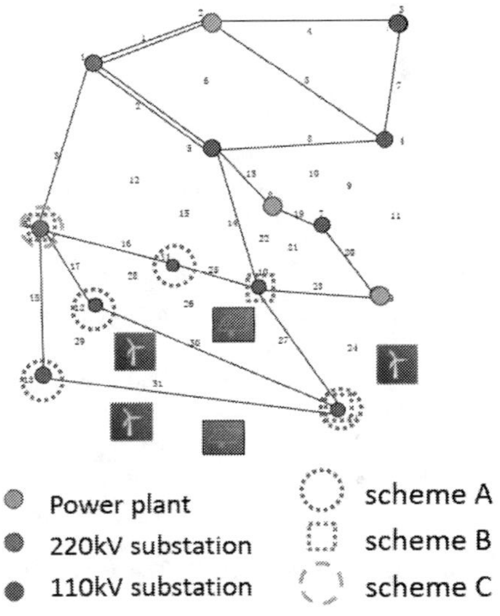

Fig. 5. The Schematic diagram of 3 planning schemes

4 Results analysis

4.1 Results

After simulating throughout of the model presented in section two, the load data at each hour of three power grid planning schemes were obtained, the efficiency results could be calculated throughout the production simulation process, the indexes of efficiency of each scheme list in Table.2. Since the clean energy increasing in power system, the power supply fluctuated randomly which caused to the load flow undulate correspondingly. As a result, the utilization ratio of the transmission line decreased in some sense compared to present efficiency, the heavy-load proportion and light-load one increased as a whole, but the index level of transformer fluctuated within a certain range; the cost of power supply came down rapidly for mass clean energy development, the abandon clean energy were different among three schemes.

Table 2. The efficiency index of planning power grid.

Index	110kV	220kV	110kV	220kV	110kV	220kV
Utilization ratio of Tran	22.5%	28.1%	26.2%	25.2%	23.8%	26.6%
Heavy load proportion of Tran	1.7%	0.56%	1.4%	0.49%	1.5%	0.52%
Light load proportion of Tran	9.1%	3.7%	5.7%	4.5%	5.9%	4.1%
Utilization ratio of Line	28.2%	25.1%	31.2%	23.1%	29.4%	24.2%
Heavy load proportion of Line	1.8%	0.43%	1.5%	0.76%	1.7%	0.53%
Light load proportion of Line	8.1%	4.1%	7.5%	6.5%	7.8%	5.4%
Cost of power supply(g/kWh)	241		242		243	
Line loss	1.4%		1.5%		1.5%	
Abandon clean energy proportion	2.1%		2.1%		2.1%	

4.2 Efficiency analysis

The comprehensive effect indicator of each scheme is proposed to reflect the improvement in terms of efficiency. It is accessed through comparing the indexes between the planning scheme and present. The formula is shown as Equation (11).

$$E_{ff} = \sum_{i=1}^{d} u_i Q_h(i) = \sum_{i=1}^{d} u_i [(k_{h-i} - k_{h-n}) / k_{h-n}] \quad (11)$$

In which, E_{ff} is the comprehensive effect indicator, Q_i is the improvement of index i which belong to the efficiency indexes in section 2.4, u is the weighting coefficient which is shown in Table.3 depending on the significance of index, k_{h-n} is the index of present grid, k_{h-1} the index of planning grid.

The results of each scheme is shown in Table.3, the E_{ff} of scheme A, B and C is positive number, it means that the comprehensive effect of those planning schemes improves compared with present one. As a result, the indicator of scheme C is maximum based on calculation, the mainly effect of scheme C is the utilization ratio and extreme efficiency ratio proportion which precedes others.

Table 3. The comprehensive effect indicator of each scheme.

Index	u	Scheme A	Scheme B	Scheme C
Q_1_Utilization ratio of Tran	0.5	-0.024	-0.016	-0.025
Q_2_Heavy load proportion of Tran	-0.3	-0.111	-0.044	-0.067
Q_3_Light load proportion of Tran	-0.3	-0.109	-0.026	-0.019
Q_4_Utilization ratio of Line	0.5	-0.015	-0.005	-0.012
Q_5_Heavy load proportion of Line	-0.3	-0.165	-0.171	-0.165
Q_6_Light load proportion of Line	-0.3	-0.045	-0.096	-0.074
Q_7_Cost of power supply(g/kWh)	-5	1.049	1.016	1.033
Q_8_Line loss	-0.5	-0.011	-0.036	-0.036
Q_9_Abandon clean energy proportion	-1	-0.400	-0.400	-0.400
E_{ff}		0.171	0.223	0.235

5 Conclusions

The efficiency of power system is closely related to the power system scale, wind plant installed capacity and the load characteristic. A production simulation model is proposed which considered those factors in power system, which includes load model, generation model, power grid model and operation scheduling. The simulation process is established to emulate the power system operation for whole year. The studied method is capable to figure out the operation condition of a planning power system rely on the current operating data, project-to-be durations and planning forecast data. The efficient indexes contains load ratio, heavy/light load proportion and generating cost, those are calculated from the simulation results, moreover, the comprehensive effect indicator is established to reflect the variation between present power grid and planning one, it could be used to access improvement the planning scheme. At last, three power grid plans to develop massive clean energy in following year, the impact of clean energy development is demonstrated by assessing the efficiency through product simulation.

Acknowledgment

This work was supported in part by the Technology Project of State Grid Corporation of China and the Technology Project of State Grid Tianjin Electric Power Company No.KJ2017-1-52.

References

1. Coal falls as gas rises. "World energy balances in 2016" [http.//www.iea.org/newsroom/news/2017/august/coal-falls-as-gas-rises-world-energy-balances-in-2016.html]
2. The editorial committee of China Electric Power Yearbook, "China Electric Power Yearbook 2013", (China Electric Power Press, Beijing, 2013)
3. H.T. Yang ; S.L. Chen, "Incorporating a multi-criteria decision procedure into the combined dynamic programming/production simulation algorithm for generation expansion planning", IEEE Transactions on Power Systems **4**(1989)
4. George A. Orfanos ; Pavlos S. Georgilakis ; Nikos D. Hatziargyriou. "Transmission Expansion Planning of Systems With Increasing Wind Power Integration", IEEE Transactions on Power Systems, **28** (2013)
5. Jeremy Hargreaves, Elaine K. Hart, Ryan Jones, Arne Olson REFLEX. "An Adapted Production Simulation Methodology for Flexible Capacity Planning", IEEE Transactions on Power Systems, **30** (2015)
6. M. Lin A.M. Breipohl; F.N. Lee, "Comparison of probabilistic production cost simulation methods", IEEE Transactions on Power Systems, **4** (1989)
7. Weiping Wu ; Zechun Hu ; Yonghua Song ; Giovanni Sansavini ; Huimiao Chen ; Xiaoshuang Chen, "Transmission Network Expansion Planning Based on Chronological Evaluation Considering Wind Power Uncertainties", IEEE Transactions on Power Systems **4** (2018)
8. A. S. D. Braga, J. T. Saraiva, "A multiyear dynamic approach for transmission expansion planning and long-term marginal costs computation", IEEE Trans. Power Syst., **20** (2005)
9. F. Bouffard, F. D. Galiana, "Stochastic security for operations planning with significant wind power generation", *IEEE Trans. Power Syst.*, **22** (2008)
10. R. Billinton, W. Wangdee, "Reliability-based transmission reinforcement planning associated with large-scale wind farms", IEEE Trans. Power Syst., **22** (2007).
11. A. Tuohy and M. O'Malley, "Pumped storage in system with very high wind pentration," Energy policy, **39** (2011)
12. B.C. Ummels, E. Pelgrum, and W.L.Kling, "Intergration of large-scale wind poer and use of energy storage in the Netherlands' electricity supply," Iet Renew. Power Generat, **2** (2008)
13. B.C. Ummels, M.E. Pelgrum, and W.L.Kling, "Impacts of wind power on thrmal generation unti commintment anddispatch", IEEE Trans. Energy, **22** (2007)
14. M.Black, V silva, and G, Strbac, "the role of the storage in intgrationg wind energy", in Proc. Int. Conf. Future Power Systems. 2005
15. N. zhang, T. Zhou, C. Duan, X. Tang, J. Huang, Z. Lu, and C. Kang, "Impact of Large –scale wind farm connecting with power grid on peak load regulation demand," (in Chinese) Power sys. Technology, **34** (2010)
16. N. zhang, C. Kang, D. S. Kirchen, Q. Xia, W. Xia, J. Huang, and Q. Zhang, "Planning Pumped storage capacity for wind power intergration", IEEE Trans. Energy, **4** (2013)
17. D. Youjie, J. D. McCalley, Vittal V." Simplification, expansion and enhancement of direct interior point algorithm for power system maximum load ability."IEEE Transactions on Power System, **15** (2000)
18. G.C.Pyo, H.W. Kang, and S. I. Moon. "A new operation method for grid-connect PV system considering voltage regulation in distribution system", IEEE power Engineering Society Gneneral Meeting, Pittsburgh.PA, July, 2008.
19. G.Notton, C. Cristofari.M. Muselli, P. Poggi,and N. Heraud. "Hourly solar irradiations estimation .from horizontal measurements to inclined data". In 1st International Symposium on Enviroment Identities and Mediterranean Area, July 2006.
20. W. Long X. Wang. and etc. "The Adequacy Discrete Analysis Based on Hierarchical Structure of Power Grid," In Chengdu . 2014 IEEE International Conference on Power System Technology Conf., pp. 1701 pp.1-7.

Wind farm layout optimization to maximize total power production

I Ulku[1], *C Alabas*-Uslu[2]

[1]Department of Industrial Engineering, Istanbul Kultur University, Istanbul, 34156, Turkey
[2]Department of Industrial Engineering, Marmara University, Istanbul, 34722, Turkey

Abstract. With the rapid increase in energy demand, renewable energy has been considered as an alternative energy resource around the world. A new nonlinear mathematical model is developed in the presented study and a comparison is obtained with a mathematical modelling study from the literature. The results superior according to power generation under a single wind direction. Jensen's wake decay model is used to state multiple wake effect among the turbines. The power production is maximized in terms of the minimization of wake effects. Mathematical formulations are represented and compared with the model from the literature. The generated layouts are presented and expected power among the `Introduction` turbines are significantly improved.

1 Introduction

Renewable energy is considered as an alternative energy resource because of its high potential to provide the required energy capacity. Continuity and sustainability opportunities of renewable energy is noticed and used as an attractive source in the developed countries. Wind power extracts the kinetic energy from the wind and converts it into electricity with the help of turbines. Turbulence or wake is generated behind the tower, while it is working. Thus, power production and the wind speed deficits of a wind turbine are reduced by another turbine's wake in the wind farm. Electricity generation with the wind farm development takes a meaningful impact with the correct estimation of the wind power. Therefore, not only the location selection, but also the placement of turbine is crucial for investment on a wind farm, in terms of maximizing power production with minimizing wake effect among the turbines.

A new nonlinear mathematical programming model is proposed where totally unimodularity property is used in the model for wind farm layout problems. The aim is to maximize total generated power while minimizing the wake effect among the located turbines. An interaction matrix is generated priori as presented in Turner et al. [1] and several number of wind turbines are tested. Wind farms are expensive investments and improvements in the investment cost will be quite valuable. The objective function where total power produced is nonlinear and binary variables of wind turbine locations make the model to be complex. In the presented model, the totally unimodularity property is introduced. Thus, the optimum turbine locations are binary without using integer programming model and generates higher total power production for a predetermined turbine numbers.

In Section 2, literature review is introduced, and nonlinear mathematical programming model is represented in section 3. The performance of the model, conclusions and future study details are is given in section 4 and 5, respectively.

2 Literature Review

Jensen in 1983 [2] has developed one of the most popular wake models. Jensen model is based on the distance behind the rotor and it is a linear expanding wake with the velocity deficit which is widely used in the literature. Figure 1 illustrates a single wind turbine wake effect. The wind blows with a speed u_0 from the left and passes through the wind turbine 1 with the rotor radius r. Wind turbine 2 is affected from the wake created by the turbine 1. Genetic algorithm is used mostly in the literature as heuristics to solve wind farm layout problems [3-5]. First model based on genetic algorithm is introduced by Mosetti et al. [3] to find optimal placement of the wind turbines. Jensen's wake decay model is used to analyse the wake effect according to various wind speeds and directions. A test case is used by dividing a square into 100 square cells where possible turbine locations can be arranged. Grady et al. [4] developed a genetic algorithm to find optimal placement of turbines, where the number of installed turbines are limited. A Monte Carlo simulation method determines optimal turbine placement in the study of Marmidis et al. [5]. Turner et al. [1] presented two mathematical models to determine the efficiency between them. First model is a quadratic-integer model and second model is a linear-integer programming model. With these models an interaction matrix is prepared priori, to determine wake effects on each possible wind turbine location.

A set of possible turbine locations are (i,j) pairs for all i,j ∈ I where I={1,...,N} is the index set. Distance between turbines at position (i,j) and (k,l) for i,j,k,l ∈ I is represented as D_{ijkl}. The wake radius r_1 and the wake enlargement, α are calculated in equation (1) and (2), respectively. Coefficients z represents hub height and z_0 indicates surface roughness [2].

$$r_1 = \alpha \cdot D_{ijkl} + r \quad (1)$$

$$\alpha = \frac{0.5}{ln\frac{z}{z_0}} \quad (2)$$

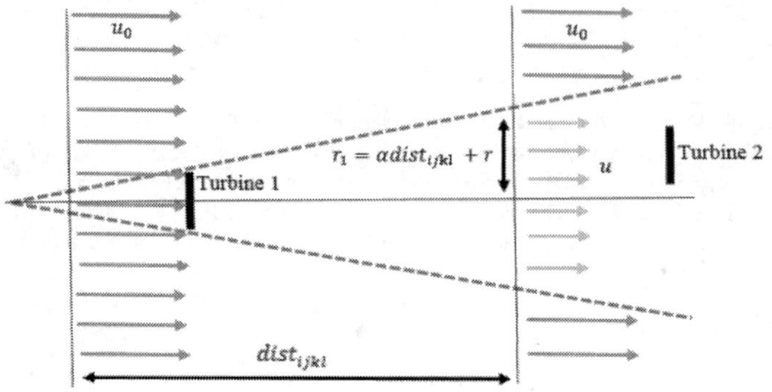

Figure 1. Wake effect representation

Wind speed, u_{ij} at position (i,j) is determined by equation (3) according to Jensen's wake model as Vd_{ijkl} is the velocity deficit in the wind speed due to the turbine located at (k,l) and u_0 is the ambient wind speed. Equation (4) is used to calculate Vd_{ijkl} by using rotor radius r, distance D_{ijkl} and axial induction factor a which is dependence of thrust coefficient, C_T as given by equation (5).

$$u_{ij} = u_0(1 - Vd_{ijkl}) \quad (3)$$

$$Vd_{ijkl} = \frac{2 \cdot a}{\left(1 + \frac{\alpha \cdot D_{ijkl}}{r}\right)^2} \quad (4)$$

$$a = 0.5\left(1 - \sqrt{1 - C_T}\right) \quad (5)$$

Multiple wake effects occur when a turbine is located at position (i,j) and affected by more than one turbine's wake. Thus, sum of the squares of individual velocity deficits generated by each wake is equal to $Vdef_{ij}$ as shown in equation (6) [2]. Equation (7) represents the remaining velocity u_{ij} by a turbine at (i,j).

$$Vdef_{ij} = \sqrt{\sum_{i=1}^{N}\sum_{j=1}^{N}\sum_{k=1}^{N}\sum_{l=1}^{N} Vd_{ijkl}^{\,2}} \quad (6)$$

$$u_{ij} = u_0 \cdot \left[1 - Vdef_{ij}\right] \quad (7)$$

3 Mathematical Programming Framework

In a wind farm layout wake effect has a key factor on the total power production since, the wind speed is decreased by individual wind turbines as explained in Section 2. A new nonlinear mathematical model is proposed under a single wind direction for the wind farm layout problem. An available wind farm is represented with a grid of size N^2 in which each cell (i,j) corresponds a possible turbine location and the number of turbines, TN, to be placed in the wind farm is known in advance. Interaction matrix of size $(N^2)x(N^2)$ is created with all possible multiple wake effects. The turbines are assumed that can turn to face at any given wind direction. Each possible turbine location has $J - 1$ interaction between itself and other possible turbine locations. A decision variable for wind farm layout problem is as follows:

$$x_{ij} = \begin{cases} 1, & \text{if there is a turbine at position (i, j) for } i,j = 1, \dots, N \\ 0, & \text{otherwise} \end{cases}$$

Since wind turbine efficiency, 40%, and hub radius, 20 m, are used in the related literature [7] the coefficient of 0.3 in equations (8) is used. To maximize total power production where the power produced by individual wind turbines (i,j) is indicated with equation (10) for wind farm layout problem. TP, total sum of the power is shown in equation (9) that is produced by all turbines located at positions (i,j). The nonlinear integer mathematical model of wind farm layout problem, is given by equations 10-12. Total number of turbines are known and satisfied with equation (11) since binary conditions of variables are represented in equation (12).

$$P_{ij} = (0.3) \cdot u_{ij}^3 \tag{8}$$

$$TP = \left\{ \sum_{i=1}^{N} \sum_{j=1}^{N} P_{ij} \qquad for\ x_{ij} = 1, \forall i,j \right\} \tag{9}$$

$$Max \sum_{i=1}^{N} \sum_{j=1}^{N} P_{ij} \cdot x_{ij} \tag{10}$$

$$\sum_{i=1}^{N} \sum_{j=1}^{N} x_{ij} = TN \tag{11}$$

$$x_{ij} \in (0,1) \qquad \forall i,j \tag{12}$$

There are some obstacles during the solution procedure of the wind farm layout. Firstly, due to the nonlinearity, the equation (10) is a main difficulty to solve the layout problem. Turner et al. [1] proposed a linearized integer programming model by minimization of total velocity deficits to handle this difficulty. Secondly, to locate a turbine or not for each possible turbine locations in a grid with size 100 consists of a huge search space. Therefore, the binary constraints in equation (12) are replaced with equation (13) with the help of totally unimodular matrix property [6].

$$0 \le x_{ij} \le 1 \text{ for } \forall i,j \tag{13}$$

Although the constraints does not take 0-1 binary values, an optimum layout to the problem is also satisfied as the binary requirements. With this relaxed form of the problem, search space is decreased and proposed nonlinear integer model with equations (10), (11) and (13) is solved simpler than the model through equations (10)-(12). The quadratic integer model of Turner et al. [1] has the following objective function where sum of the multiple wake effects are minimized as in equation (14). The quadratic integer model of Turner et al. [1] is represented with equations (14), (11), and (12) and compared with the proposed nonlinear integer model in this study.

$$Min \sum_{i=1}^{N} \sum_{j=1}^{N} \sum_{k=1}^{N} \sum_{l=1}^{N} x_{ij} * x_{kl} * Vdef_{ij} \tag{14}$$

4 Computational Results

A set of test problems are generated to represent the performance of the proposed nonlinear model for single wind direction from north to south. The parameters in table 1 are used in this study. An interaction matrix is prepared priori in order to determine velocity deficits of each possible turbine locations.

Table 1. Input parameters for the cases.

Grid size	100 square cells
Cell size	$10r_0$ (200 m x 200 m)
Wind speed (u_0)	12 m/s
Rotor radius (r_0)	20 m
Hub height (z_0)	60 m

Thrust coefficient (C_T)	0.88	
Entrainment constant (α)	0.1	
Power (P_{ij})	$P_{ij} = (0.3) * u_{ij}^3$	Ref. [3]

In the following subsection, the obtained expected power results from the proposed nonlinear model are compared with the quadratic integer model by Turner et al. [1].

4.1 Comparison of Non-Linear Model with Quadratic Integer Model

Both non-linear model and quadratic integer model are performed in LINGO [8] on a 1.70 GHz Intel Core i5 notebook with 6 GB of RAM with an upper limit of 2 hours on the run time. Thus, total power results of non-linear model are the lower bounds on the exact optimum power for the test cases of wind farm layout problem. Generated total power results by non-linear model and quadratic integer model are introduced in Table 2 for several number of turbines considering single wind direction from north to south. Although for 10 and 20 turbines both models obtain the same total power, non-linear model has better power output than quadratic integer model for 30 and more turbine layout. Apparently, as the number of turbines increases the performance of non-linear model proposed in this study also increases comparing the quadratic integer model. Since the objective function of the non-linear model is to maximize total power generated, this model is more successful than the quadratic model which minimizes the total velocity deficits. Although, minimization of total velocity deficits is necessary to generate more power, this study shows that if the objective function is selected as maximization of total power instead of minimizing velocity deficits, the resulting wind farm layouts will be more efficient.

Table 2. Single Wind Direction from north to south

Turbine Number	Turner et al. Power (kW)	Non-Linear Model Power (kW)
10	**5760**	**5760**
20	**11364.9716**	**11364.9716**
30	13976.48032	**16163.17394**
40	16390.31483	**19609.61508**
50	16302.32057	**22644.51072**

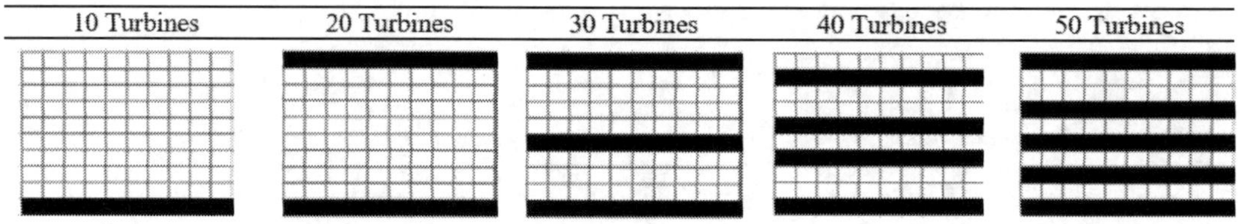

Figure 2. Non-Linear Model Layouts for a Single Wind Direction

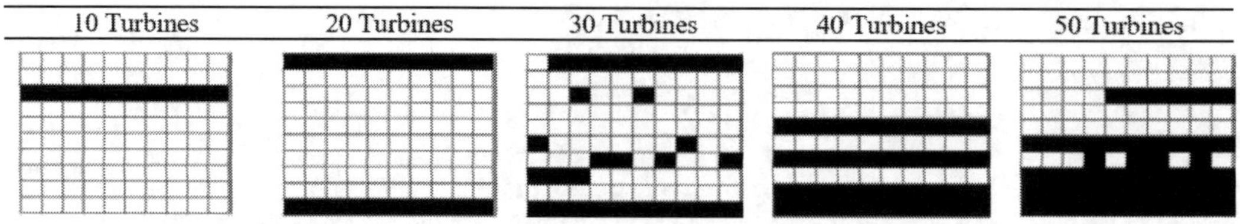

Figure 3. Quadratic Integer Model Layouts for a Single Wind Direction

Figure 2 and Figure 3 gives the optimal wind farm layouts of non-linear model and quadratic integer model for the turbine numbers from 10 to 50 turbines under single wind direction.

5 Conclusion

A new nonlinear mathematical model is presented to solve layout of wind farm problem to maximize total power production in terms of multiple wake effects. Jensen's multiple wake decay model is used for the presented mathematical model. Due to nonlinearity of the power and wake effects calculations the researchers preferred to use heuristic methods in the literature. Layout of a wind farm is a permanent investment, thus, any improvement in the layout of a wind farm increase the generated total power. The proposed nonlinear mathematical programming model has totally unimodularity property where binary variables of wind turbine locations is relaxed and model is solved simpler. It is obvious that solving the proposed nonlinear model within a limited run time generates better lower bounds on the exact optimum power in comparison with the quadratic integer model [1]. Also, an interaction matrix is used priori as presented in Turner et al. [1] and various number of wind turbines are performed. The proposed study express that the wind farm layout problems can have optimum layouts. We are planning to extend our study by including multiple wind direction. Thus, various situations can be handled for the wind farm layout problems.

References

1 Turner SDO, Romero DA, Zhang PY, Amon CH, Chan TCY. A new mathematical programming approach to optimize wind farm layouts, 2014, *Renewable Energy*. **674-680**.

2 Jensen NO. A note on wind generator interaction *RISO National Laboratory*; 1983 November; Roskilde, DK-4000, Denmark.

3 Mosetti G, Poloni C, Daviacco B. Optimization of wind turbine positioning in large windfarms by means of a genetic algorithm. 1994, *J Wind Eng Ind Aerod*. 51:**105-116.**

4 Grady S, Hussaini M, Abdullah M. Placement of wind turbines using genetic algorithms. 2005, *Renewable Energy*. 30: **259 - 270.**

5 Marmidis G, Lazarou S, Pyrgioti E. Optimal placement of wind turbines in a wind park using monte carlo simulation, 2008, *Renewable Energy*. 33: **1455 - 1460.**

6 Wolsey LA. *Integer Programming*, Wiley, 1998.

7 Mittal A. Optimization of the Layout of Large Wind Farms using a Genetic Algorithm. Department of Mechanical and Aerospace Engineering, 2010, Case Western Reserve University, Cleveland.

8 Extended LINGO/Win 64. Release 17.0.65. Lindo System Inc. Chicago, September 2017.

Physical tests in the selection of electrical lineman depending on the cardiac rhythm when working at height for the first time

Ernesto A. Abril[1], *Ernesto* M. Abril[1], *Diana* D. Marín[1], *Leonardo* M. Abril[2]

[1]Facultad de Ciencias de la Ingeniería y Aplicadas, Universidad Técnica de Cotopaxi , Latacunga-Ecuador
Departamento de Ciencias Económicas Administrativas y de Comercio
[2]Universidad de las Fuerzas Armadas ESPE, Sangolquí-Ecuador

Abstract. In the present work of investigation, the pulse cardiac is monitoring with a heart rate sensor. All data record is obtained of a healthy person group, five junior and one expert electric lineman, Pulse cardiac let show the behavior of the heart when responding to physical effort and stress situation. So that we relate the required physical effort evidenced by a calorific map, the condition of stress due to the fear of falling and the variability of the cardiac pulse. Study is presented in stages. In the first stage, it is possible to demonstrate the increase of the cardiac pulse due to the effort. In the second stage, the increase of the cardiac pulse due to a condition of stress or fear is presented. Finally, in the third stage, the cardiac pulse increases proportionally. Also the cardiac pulse is monitored when descending the concrete post. A variability of the cardiac rhythm between the participants is obtained. It checks that the heart rate after an effort or stress condition becomes unpredictable for different people.

1. Introduction

In the electrical sector, the cardiological effects produced by work on electrical towers and concrete poles have not been studied. The heart rate in the electrical linemen allows to determine if a danger will occur due to its nervous state. This is because when working in medium voltage, an error can make the difference between life and death.

On the other hand, the cardiological information of the prospects to begin in this high-risk work is of vital importance. This is because the electric lineman, when presenting a state of anxiety, would cause him to make an error in the medium voltage lines. Anxiety is clearly observed in cardiac pulse curves with high pulse-to-pulse variability. Showing steep slopes.

The electric lineman is in charge of maintaining the high voltage towers and the distribution lines. Infrastructure that allows the transport of energy from generators to consumers. Manuals for the selection of personnel in the electrical line work post indicate that it is necessary to monitor the cardiac pulse of the prospects. The pulse must be kept below the maximum permissible. [1]

In the studies the physical and psychological behavior are presented. The recommendations are presented before, during and after performing the physical tests. This is done to get the best results.

For the physical tests is recommended, that each of the participants avoid the consumption of tobacco and perform aerobic and walking activities. [1]

Cardiac pulse monitoring is essential to analyze the behavior of the heart when performing physical activities and stress situations. Most cardiological studies require the acquisition and analysis of heart signals. For this reason, heart rate sensors and ergometers are used.

This monitoring of cardiac pulse has served to validate physical activity questionnaires. Since it allowed to demonstrate significant correlations between the cardiac pulse and physical activities in different levels of intensity. The study method consisted of analyzing a multi-ethnic sample from New Zealand, which consisted of people from the Pacific, Maori and Europeans.

The multi ethnic people were monitored during a period of three consecutive days, to later associate the cardiac pulse to the different activities of a daily day. Ultimately the study aimed to establish a guide with suggestions of physical activities to maintain a good state of health in the local population.

On the other hand, the statistical analysis is based on the recommendations by the American College of Sports Medicine, which establishes that for moderate exercises there will be a cardiac pulse reserve of 40% to 50% and for high intensity exercises of 60-70% and the maximum cardiac pulse is:

$$HR_{max}=220\text{-age} \quad (1)$$

In which HR_{max} is the maximum cardiac pulse depending on the age of the person. [2]

The cardiac pulse has been studied for many years. It is because there are many devices to analyze these signals such as electrocardiograms. The device records the frequency and variability of the cardiac pulse.

 https://doi.org/10.1051/e3sconf/20185703006

In the psychological aspect the heart rate as a measure of stress or anxiety. It has been noted that depending on the state of mind and under extreme conditions a person can increase or decrease their heart rate. However, this variability cannot be greater than unity.

Experiments were performed in which the heart rate was checked before, during (exposure of the individual to a stress condition) and afterwards. This condition consisted in keeping one foot on ice. From the experiment it was observed that in a stress condition the person accelerates his heart rate.

In another experiment, an individual with good physical condition was observed. It consisted in that he performs push-ups to the point of exhaustion, then the data obtained of cardiac pulse in the post-stress stage are unstable and unpredictable. [3]

In other experiment shows that the pulse and heart rate varies, depending on whether the person is asleep or awake. Four people were employed in good health. These people would make a trip of 5 hours. During the trip they had to press a button in case of feeling drowsy. Throughout the experiment people were being monitored.

The data extracted indicated a 10-20% decrease in the cardiac pulse and a 10% increase in the variability of the cardiac pulse. [4]

The cardiac pulse has also been linked to phobias. In some experiments people with spider phobia have been used. In the subjects, the cardiac pulse was monitored before and after some tests. The tests consisted of touching a spider's web to touching a living spider. The data from the experiment show an increase in the cardiac pulse when the subjects face a phobia. [5]

Experiments have been developed that involve people in good physical condition. The people were subjected to continuous and constant bicycle exercises. The cardiac pulse was analyzed as a function of time and presented two components. [6] The first component presented an abrupt increase and in the second component the cardiac pulse stabilized with a cardiac pulse close to the maximum HRmax. [2] [6]

The results were obtained in the city of Ambato, Ecuador in South America. The city is located at a height of 2700 meters above sea level. The electric distribution lines expand to areas far from the cities, for this reason the work sectors are above 3000 meters above sea level. These working conditions require a greater physical effort and therefore a greater number of cardiac pulses by the electric line.

Electrical linemen are equipped with foot couplings, seat belts and helmets. The foot couplings are hooked on the concrete poles in each step that is given to the ascent. The weight of the equipment is 30 [kg]. Before the start of the physical tests, the participants were weighed, if their weight exceeded 285 pounds they were not taken into account for the tests. Smoking and eating were not allowed in the course of the tests, as was the use of cell phones. [7]

The study presents two parts of observation in the first one show the caloric corporal points in which greater effort is made, in the second part the curves of the cardiac pulse of each one of the participants are obtained. Finally the two parts of the experiment are analyzed.

2. Development

2.1 Electrical lineman thermal body map

Initially, the body thermal map was observed. An increase in temperature is observed in the areas in which greater effort is required Figure 1. The progress of the electric lineman in the ascent to the pole is observed.

Figure 1. Start climbing

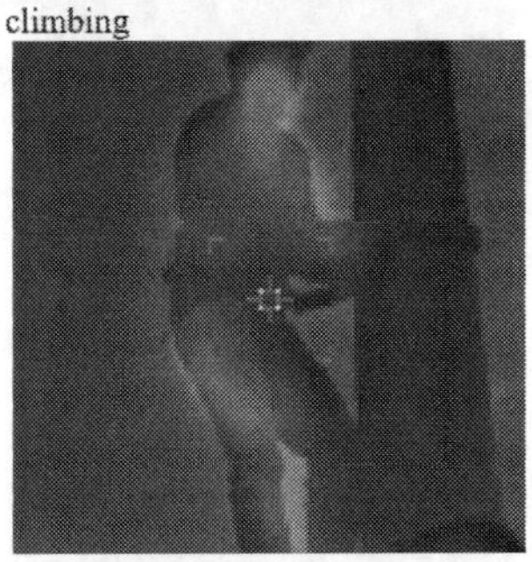

Figure 2. Middle climbing

Figure 3. Finish climbing

The images show a concentration of heat in the arms, chest and legs. Figure 2. Thermal radiation protrudes above the clothes. The body radiation emitted by the face of the person is discarded.

At the end of the ascent, the highest concentration of heat is found in the thighs. This is due to the greater effort required. Figure 3.

2.2 Cardiac pulse monitoring rise.

The heart rate was recorded using a sensor with electrodes. The sensor has communication by bluetooh to record the data..

Six people participated in the experiment. A person has experience in this activity. The other 5 people carry out this activity for the first time.

The cardiac pulse curves of Figures 1, 2, 3, 4, 5and 6 are from junior linemen as they ascend.

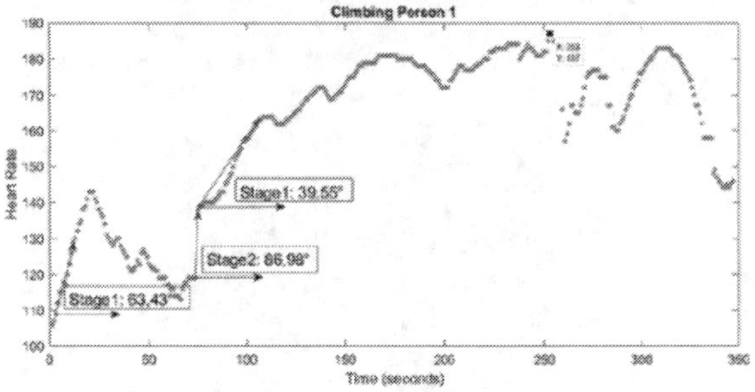

Figure 4. Person 1 Climbing by stages.

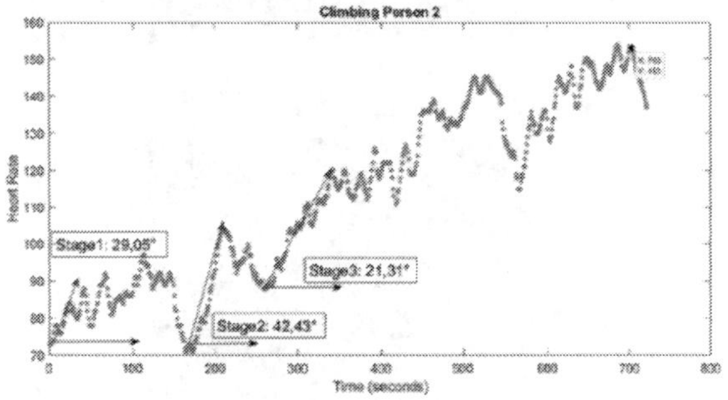

Figure 5. Person 2 Climbing by stages.

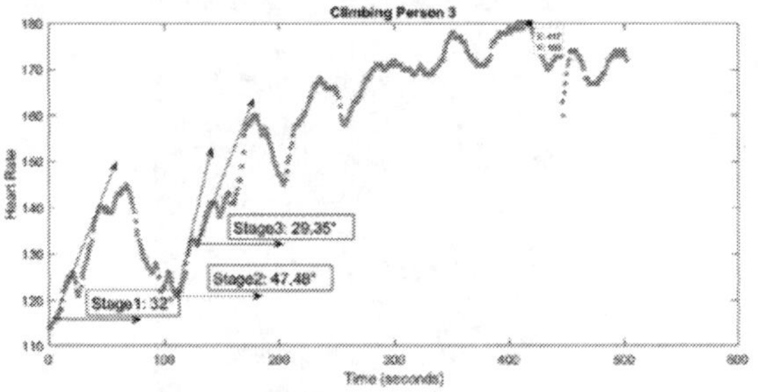

Figure 6. Person 3 Climbing by stages

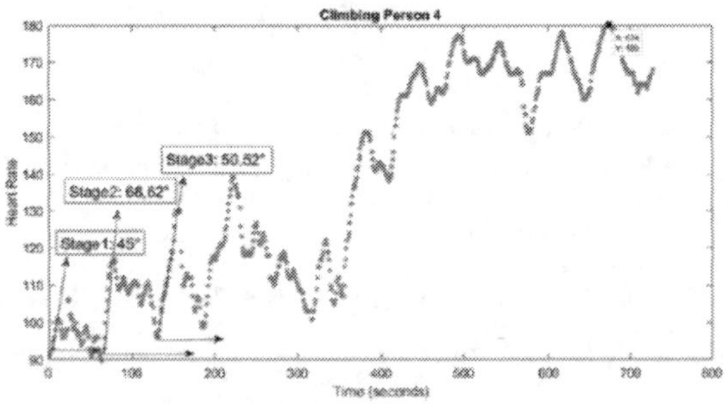

Figure 7. Person 4 Climbing by stages

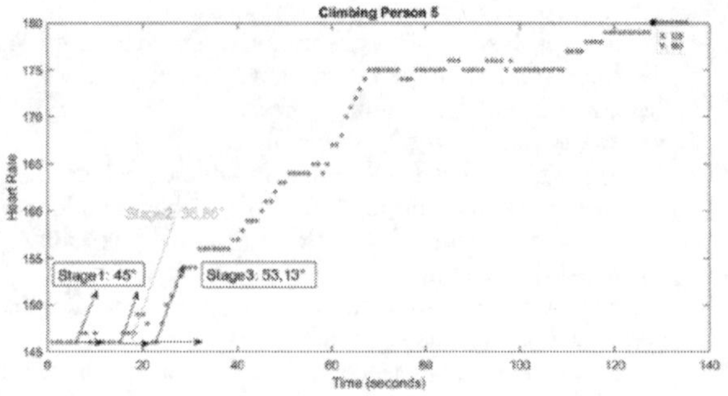

Figure 8. Person 5 Climbing by stages

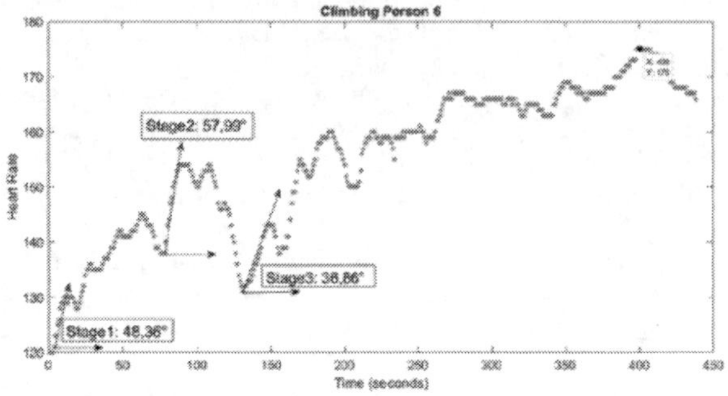

Figure 9. Person 6 Climbing by stages

2.3 Cardiac pulse monitoring when descending.

The maximum heart rate [1] was fixed according to the average age of the group of people observed. The average age is 30 years.

$$HR_{max}=220\text{-}age \qquad (1)$$

$$HR_{max}=220-30=190$$

At the end of the climb and start the descent the following curves are presented. The blue line is the maximum cardiac frequency that is permissible. Figure 10, 11, 12, 13, 14 and 15.

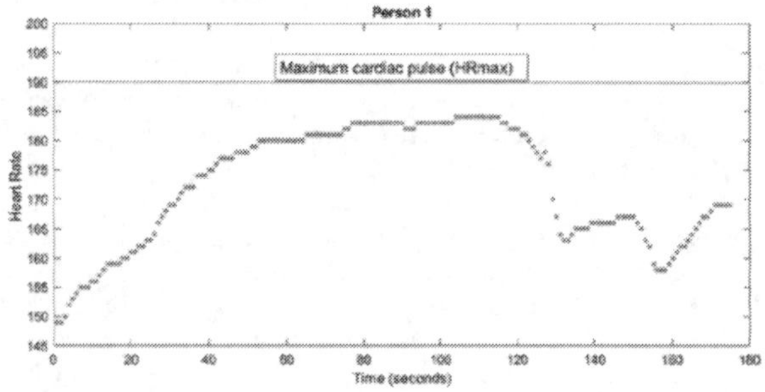

Figure 10. Person 1 Descending

https://doi.org/10.1051/e3sconf/20185703006

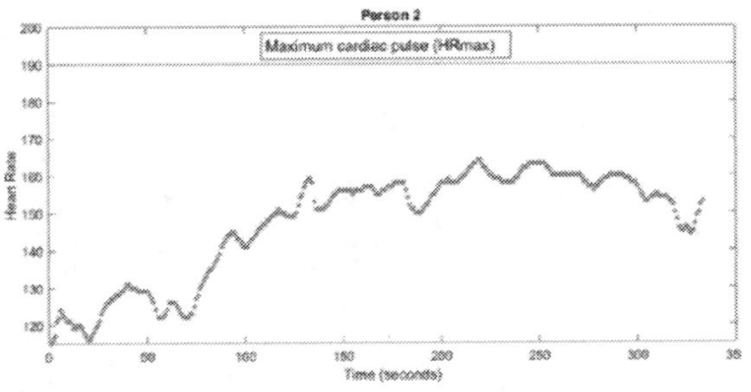

Figure 11. Person 2 Descending

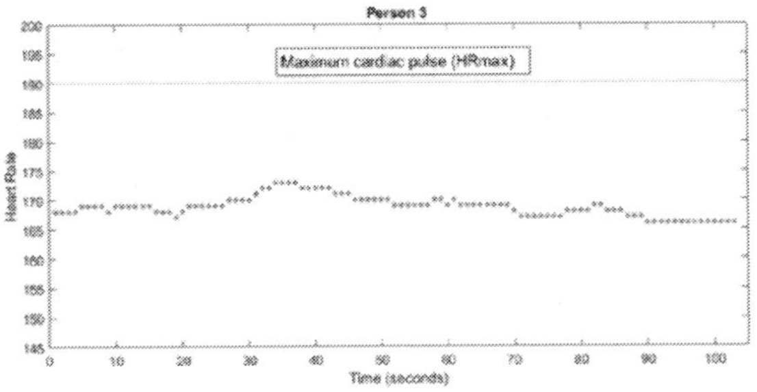

Figure 12. Person 3 Descending

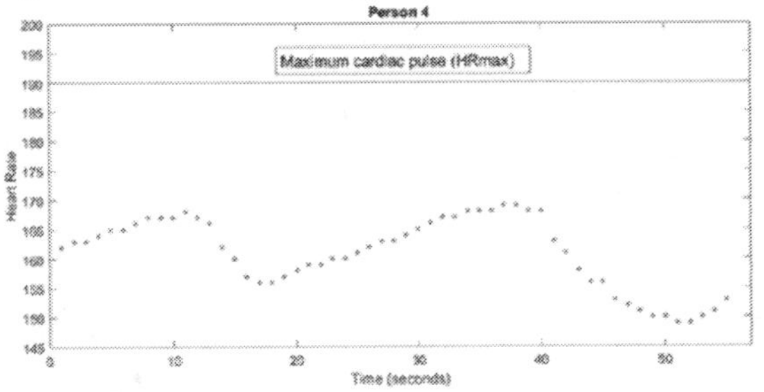

Figure 13. Person 4 Descending

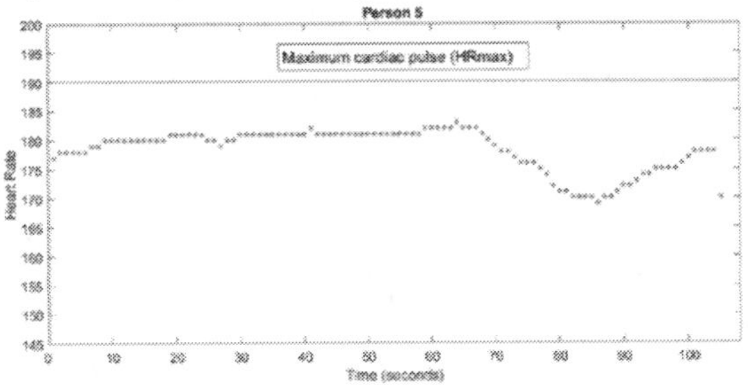

Figure 14. Person 5 Descending

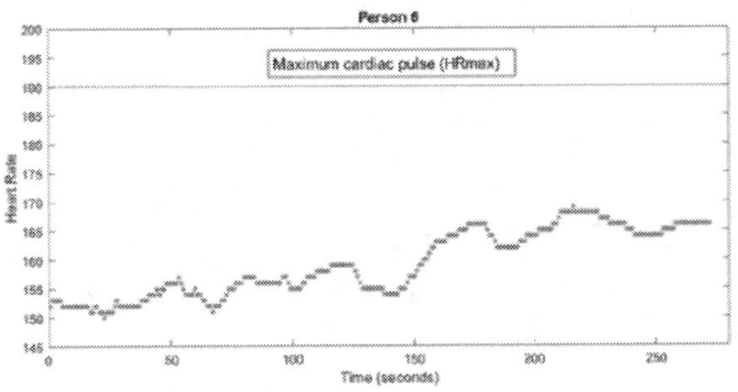

Figure 15. Person 6 Descending

3. ANALYSIS OF RESULTS

According to each of 6 images the column of stage 1 present the first step, the stage 2 is the second step and stage 3 represent steps for climbing all concrete post. Table 1.

In stage 1 it gives the first step on the concrete post but still keeps the other foot on the ground. This is what reflects the effort that this first step requires. In stage 2, he gives the second step on the concrete post. The person loses contact with the ground and for this reason is in a state of anxiety. Due to the fear of falling. In stage 3, the vertical ascent to the concrete post begins. For this reason every step that the union of fear and effort presents, maintaining a constant and continuous increase up to the limit of the cardiac frequency.

Table 1. Climbing with data angular slope by stages

	Angular slope climbing		
	Stage 1	Stage 2	Stage 3
Person 1	63.43°	86,98°	39,55°
Person 2	29,05°	42,43°	21,31°
Person 3	32°	47,48°	29,35°
Person 4	45°	68,62°	50,52°
Person 5	45°	36,86°	53,13°
Person 6	48,36°	57,99°	36,86°

Results of Table 1, Of the 6 people monitored, the fifth one has experience in this type of work. This is evidenced by the lower cardiac pulse in the second stage compared to the other participants. The rest of the participants carry out this activity for the first time.

The inexperienced participants have a greater heart rate in the second step they take when climbing the concrete post. This is due to the fear of falling since from there they fall off the ground.

However, in the third stage, the increase of the cardiac pulse in both the experienced participant and those without experience, maintain a continuous and constant increase..

When starting with the descent of the concrete post, the cardiac pulse in each one of the participants becomes unpredictable and variable.

4. Conclusions

The cardiac pulse tends to increase when a person is in a state of stress and stress. The heart rate allows to differentiate each one of these states. Coinciding with greater variability in the second stage effect of the state of stress and fear. However, in people with experience in this type of work the heart rate decreased with respect to the first stage, denoting the absence of the effects of stress. When concluding with the ascent, the cardiac rhythm is difficult to predict. This is because the physical state and physiology of each of the participants is different. Causing different forms of curve in each of them.

5. Acknowledgment

Our sincere thanks to Universidad Técnica de Cotopaxi for collaborating with the present project with the logistics for the respective presentation and Msc. Idalia Pacheco for the eloquent administrative management.

6. References

1. Pacific Gas and Electric Company, "Post-Offer Physical Assessment – Lineman Line of Progression", PG&E Corporation, Available: www.pge.com [Accessed Febrary. 14, 2017]
2. Karen L. Moy, Robert K. Scragg, Grant McLean, and Harriette Carr, "The New Zealand Physical Activity Questionnaires: Validation by Heart-Rate Monitoring in a Multiethnic Population", Journal, Physical Activity & Health vol 5, pp. 45–61, 2008.
3. J. Alan Burdick and John T. Scarbrough, "Heart Rate and Heart Rate Variability an Attempt to Clarify", Journal, Perceptual and Motor Skills vol 26, pp. 1047–1053, 1968.
4. Sang-Ho Jo, Jinmyung Kim and Dongkyu Kim, "Heart Rate and Heart Rate Variability Change While Falling Asleep During Driving a Car", Journal, American College of Cardiology vol 70, pp. 79, 2017.
5. C. B. Taylor, "Heart-Rate Changes in Improved Spider-Phobic Patients", Journal, Psychological Reports, vol 41, pp. 667-671, 1977.
6. W. Michael Felts, Stephen Crouse and Mark Brunetz "Influence of Aerobic Fitness on Ratings of Perceived Exertion during Light to Moderate Exercise", Journal, Perceptual and Motor Skills , pp. 671-676, 1988.
7. Ediscon International Company, "Lineman Physical Performance Test Battery", Southern California Edison, Available: www.edison.com [Accessed Febrary. 15, 2017]

A New Distribution System Performance Approach to the Switch Allocation Problem Under Smart Grid Framework

Hatem Zaki[1], *R.A.* Swief[2], *T.S.* Abdel-Salam[2], *M.A.M.* Mostafa[2]

[1]Distribution System Planning, BC Hydro, Vancouver BC, Canada
[2]Dept. of Electrical Power & Machines, Ain Shams University, Cairo, Egypt

Abstract. This paper proposes a new vision of the Distribution System Switch Allocation problem considering new system performance measures. The mathematical model has been rebuilt with a new aggregated multi-objective formula, minimizing a newly developed performance index while achieving minimum annual energy lost. A new practical weighted combined system performance index, consisting of Reliability, Resiliency and Vulnerability, is applied and tested to be used by utilities replacing the common simple reliability index combination. The new model uses mixed integer design variables to determine the number, location and status of switches. A set of eight logical and technical constraints was applied to provide the best description of the real existing system constraints. A new algorithm of checking the system radial topology is also applied to the problem. The problem was solved using the Genetic Algorithm and was tested on a 54-bus real distribution test system, deemed more complicated than the test systems found in literature, to demonstrate its validity and effectiveness in real life systems.

1 Introduction

In today's decision making, under Smart Grid (SG) approaches, system performance is a vital characteristic of the Distribution System (DS) Planning problem and must be combined as part of the objectives when making such significant decision [1].

Switch Allocation (SA) is a crucial and attractive topic to power DS planners and operators. As presented by past researches, presence of switches in DS is nowadays essential, for the following three main goals:

(1) To sectionalize radial distribution feeders, for the purpose of limiting outage durations by limiting the number of customers out of service during faults [2],

(2) To tie more than one feeder for the purpose of creating open loop topologies to facilitate power restoration during contingencies [3], and

(3) To reconfigure the DS for minimization of active power (and energy) losses in the DS during normal operations [4].

While the first and the second goals play a role only in improving the System Performance (SP), the third goal will impact the System Operational Flexibility (SOF) in addition to overall system financial performance.

Strategically selecting the location of switches in the DS and, perhaps automating some or all of them, plays a role in maximizing the value of certain objectives [5]. Commonly used objectives, in past research, often included maximization of reliability [6], minimization of switch capital and running costs [7], minimization of cost of outages [8], minimization of total Expected Energy Not Served (EENS) and minimization of system losses [4]. Combinations of these objectives have been used in the past as the main interest of researchers around the subject of SA in DSs. Even though these past researches used EENS as an objective to express the SP, it seems that utilities interest is more focused towards common performance indices. These common indices include System Average Interruption Frequency Index (SAIFI), System Average Interruption Duration Index (SAIDI), percentage of Customers Experiencing Multiple Interruptions of 4 or more (CEMI-4), and Customers Experiencing Lengthy Interruption Durations of 6 or more hours (CELID-6) [9].

In this paper, a new formulation of the DS model for the purpose of SA including a new combinatorial performance index is proposed and demonstrated using test cases. The proposed index addresses utility concerns and is proven to be a good indicator of the system performance. The paper neither discusses, nor proposes the automation of switches in the DS but rather focuses on presenting the proposed model with its application and results.

Previous research on finding the optimal placement and the number of switches in DS involved several solution methodologies. These methodologies included analytical hierarchical process decision making algorithm [10], heuristics [11], immune [12], reactive tabu search [6] or decomposition approach [13]. Even though good results have been obtained from these methodologies, they only applied a smaller number of constraints. These constraints are necessary to make the system simulate the real-world applications.

The solution methodology used in this paper is the Genetic Algorithm (GA) with modifications necessary to obtain the best solution of the SA problem. A list of eight constraints has been applied and divided into two main groups to better simulate the real-world systems and make the proposed methodology useful for planners and operators. A new approach to one of the constraints has also been applied to ensure the usefulness of the obtained results.

The paper is divided into six sections. In addition to this introduction, the new model description is presented in section II. The mathematical formulation representing the proposed model is discussed in section III and the solution algorithm is described in section IV. Test cases and conclusion are presented in sections V and VI respectively.

2 Problem description

On the roadway of using multiple objectives as targets for locating switches, this paper proposes the use of an aggregated sum of two components: (*A*) a new combinatorial performance index formed of three common performance indices, and (*B*) the system annual energy losses. The goal of the problem is to minimize the aggregated sum of both objectives.

2.1 The new combinatorial performance index

Even though there is a clear need for an expressive system performance index under the smart Grid framework, neither regulators nor utilities were able to agree on a complete standard understanding of how to represent system performance. Standard reliability indices have been used and proposed by IEEE with a later improvement to include resiliency to cover expected large environmental impacts of systems [9]. Most utilities have combined these indices in one formula with weighted sums tailored to their needs [14].

Following this trend, the proposed Contingency Index (CONDEX), in this paper, is formed of three components:

i. The Unified Reliability Index (URI) – This index has been often used to express reliability of a given DS with historical outages. URI can be formed of the weighted sum of multiple known reliability indices [14]. In this paper, and for practical reasons, only four indices were used, which are SAIFI, SAIDI, CEMI-4 and CELID-6. Other indices require special equipment to obtain data which is usually not available for most utilities due to their significant costs.

ii. The System Resiliency Index (SRI) – As proposed by the IEEE guideline [9], this index is CELID-12, which reflects the percentage of customers experiencing 12 or more hours of outages. Many utilities have modified this index to include the number of years, where CELID-12 has happened, over the study period. In this paper, the new modified SRI, including the number of years, is used.

iii. System Vulnerability Index (SVI) – This is the new index presented in this work. SVI represents the ability of the DS to stay in-service during and after a massive disaster such as a massive earthquake, a one of a kind storms with destructive wind speed (not annual storms), large permanent floods, etc. To use a predictive Vulnerability index, three weighted arguments are created and selected to form the SVI. This performance index is a function of the following arguments:

- Node Distance Index (NDI) which presents the distance between each node and its source (substation in most cases).
- Node Failure Rate Index (NFRI) which presents the failure rate of each node route as linked to its source.
- Node Failure Duration Index (NFDI) which presents the failure duration of each node route as linked to its source.

SVI is then formed of the sum of the weighted values of NDI, NFRI, and NFDI.

CONDEX is hence formed of the weighted sum of URI, SRI and SVI. The presence of these weights provides enough flexibility to adjust the system configuration according to the highest priority index according to the strategy of the utility.

2.2 The system annual energy losses

In determining energy losses many researches have approximated the daily load curve to off-peak and peak averages and have taken seasonal annual averages to represent annual loads [15]. In modern systems, if digital monitoring devices are used, obtaining hourly annual loads, and hence a load factor (LF) is possible.

The Loss factor can then be determined using the common approximate formula created by Buller and Woodrow [16].

$$FLS = (1 - k) * LF + k * LF^2 \tag{1}$$

Where,

FLS *is the Loss Factor to be determined,*
k *is an imperical constant*

k can be determined using examination of circuits of a certain utility and if not known it can be assumed 0.85 [17]. The total Annual Energy Losses (AEL) can then be determined using the following formula:

$$AEL = Peak\ Power\ Loss * 8760 * FLS \tag{2}$$

The Peak Power Loss can be found from the load flow of each line section. The 8760 is the total number of service hours per year. Each load can be represented by its own annual load shape as was done in the EPRI Green Circuits project.

In this paper, annual hourly loads were used for the study systems to determine annual energy delivered. It could be safely assumed that this annual load curve is standard and that the system will follow this annual load curve every year.

AEL of a DS is a function of loads and can be calculated using load flow for each line section. FLS can be calculated using equation (1). Then AEL can be determined for each line section using equation (2). Based on this formula, the system will have a different AEL for each configuration depending on line section status. AEL can also be expanded to reflect growth in energy losses every year for all years of the study period.

The above-mentioned Objectives are subjected to a number of constraints. These constraints limit the optimum solution to a practical implementable solution, making the model as close as possible to the systems implemented in real life. These constraints are explained further in the Mathematical Formulation section.

3 Mathematical formulation

As mentioned in the Problem Description section, the objective function is formed of two parts aggregated and minimized under one representation. To avoid unit and number compromise in the objective function, the two parts of the objective function have been normalized by referring them to the maximum values in the system. While $CONDEX$ is referred to the maximum value, AEL is referred to the total energy required by the system. This way AEL becomes comparable to $CONDEX$ as it will always be less than 1.

As such the mathematical minimization problem can simply be as follows:

$$Minimize\ F = CONDEX + AEL \qquad (3)$$

Equation (3) describes the overview of the objective function. The components of the Objective function consist of two parts, in addition to the constraints. The details are described in the following subsections.

3.1 Formulation of contingency index (CONDEX)

$CONDEX$ is the first objective and its formula is as follows:

$$CONDEX = A * URI + B * SRI + C * SVI \qquad (4)$$

Where,

A *is the weighting factor of the Unified Reliability Index* (URI)
B *is the weighting factor of the System Resiliency Index* (SRI)
C *is the weighting factor of the System Vulnerability Index* (SVI)

i- URI is mathematically defined as:

$$URI = a1 * SAIFI + a2 * SAIDI + a3 * CEMI - 4 + a4 * CELID - 6 \qquad (5)$$

Where,

$a1, a2, a3, \& \ a4$ *are the weighting factors of each reliability index*
$SAIFI$ *is the reliability Index known as System Average Inerruption Frequency Index*
$SAIDI$ *is the reliability Index known as System Average Interruption Duration Index*
$CEMI - 4$ *is Customers Experiencing Multiple Interruptions of 4 or more*
$CELID - 6$ *is percentage of number of Customers Experiencing Lengthy Interruption Durations of 6 hours or more*

ii- SRI is mathematically defined as:

$$SRI = CELID - 12 + Nyrs \qquad (6)$$

Where,

$CELID - 12$ *is the percentage of Customers Experiencing Lengthy Interruption Durations of 12 hours or more for a given number of years*
$Nyrs$ *is the ratio of number of years there has been 12 hours or more outages during the span of given number of years* $CELID - 12$ *has been applied*

It is common to use 5 years for most cases as the ultimate number of years for system resiliency measurement.

iii- SVI is mathematically defined as:

$$SVI = c1 * NDI + c2 * NFRI + c3 * NFDI \qquad (7)$$

Where,

$c1, c2, c3$ *are the weighting factors of each vulnarability index*
NDI, $NFRI$ and $NFDI$ are as previously defined in the Problem Description section.

3.2 Formulation of the annual energy lost (AEL)

As indicated in the Problem Description section AEL can be calculated for each individual line section. Therefore, the system AEL can be as follows:

$$AEL = \sum_{t=1}^{T} \frac{1}{E_{st}} \sum_{i=1}^{m} P_{Li}^{(t)} * 8760 * FLS \qquad (8)$$

Where,

m *is the total number of line sections*
T *is the total number of study period in years*
$P_{Li}^{(t)}$ *is the Peak Power Lost of section i in year t*
E_{st} *is the Energy required by the system in year t*
FLS is calculated from the formula shown in equation (1) as mentioned above.

3.3 Problem constraints

The above-mentioned objectives are subjected to a number of constraints. These constraints consist of two sets of constraints: (a) Logical and, (b) Technical as follows:

(a) Logical Constraints:

i. Radial Structure – After isolating all power sources except substations, from the system under study, this constraint algorithm then performs two checks on the given system: (1) Internal Loops and, (2) connectivity to supply. Fig. 1 shows an overview of the proposed radial structure algorithm.

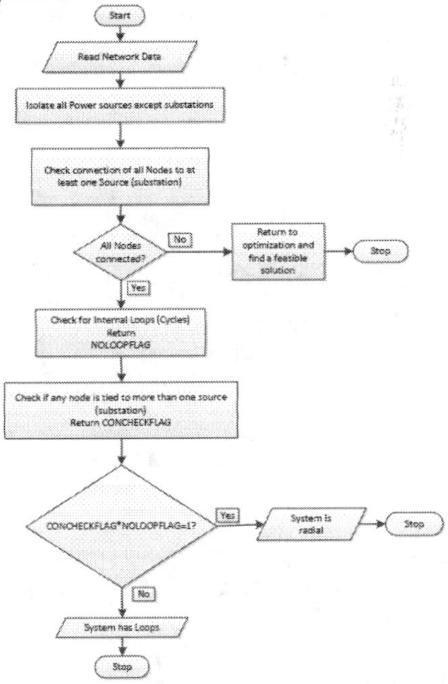

https://doi.org/10.1051/e3sconf/20185703007

Fig. 1. Radial Structure Checking Algorithm overview.

The algorithm returns a flag $(Radialflag)$, which returns 1 if the system is radial or 0 if the system contains loops or supplied by multiple substations.

ii. Node Connectivity – This is also implemented by an algorithm returning a flag called $(concheck)$.

iii. Limit number of Normally Opened (NO) switches per feeder. This constraint can also be replaced with a cost limiting constraint to limit the total costs of switches being proposed in the system.

(b) Technical Constraints:

i. Voltage Limits Constraints – Voltages of all system nodes must be within standard ranges between a minimum value and maximum value as shown in inequality (9).

$$V_{min} \le V_i \le V_{max}\,, i = 1,2,3 \dots, n \qquad (9)$$

Where,

n *is the total number of nodes not including source nodes*

V_{min} *and* V_{max} *are the standard allowable voltage limits*

V_i *is the voltage of node i*

iii. Substation Active Power Balance constraint assuming no reactive power limit, shown by equation (10)

$$\sum_{i=1}^{n-sub} P_i \le P_{max}^{(S)} \qquad (10)$$

Where,

$n - sub$ *is the total number of nodes connected to a substation S*

$P_{max}^{(S)}$ *is the maximum Active Power rating of substation S*

P_i *is the active power of node i connected to substation S*

iv. Distributed Generation Power Balance constraint assuming no reactive power limit, shown by equation (11)

$$\sum_{j=1}^{ng} PG_j \le PG_{max}^{(G)} \qquad (11)$$

Where,

ng *is the total number of Line sections emerging from generator G*

$PG_{max}^{(G)}$ *is the maximum Active Power rating of generator G*

PG_j *is the active power of Line section j emerging from generator G*

v. Line section current thermal limits, shown in inequality (12)

$$I_i \le I_{max} \qquad (12)$$

Where,

m *is the total number of line sections*

I_{max} *is the line section conductor allowable current thermal limit*

I_i *is the line section current flow*

In past research, only a few of these constraints were applied to the SA problem. In this paper, all the above-mentioned constraints were successfully applied producing good results.

4 Solution methodology

The solution methodology of the SA problem, modeled in this paper is a robust approach to avoid long processing times, often found in past methodologies. The algorithm can be summarized in the following steps:

Step 1 – starts by determining if a line section is a lateral. If the line section is not a lateral it is a candidate for either a NO or a Normally Closed (NC) switch. In this step, laterals are also, reduced to their equivalent nodes. The system model is then rebuilt without laterals.

Step 2 – This step is similar to the reconfiguration problem solution. After determining candidates of system switches, the algorithm determines the best configuration of the DS using the identified two objectives mentioned in equations (4) and (8) including all logical and technical constraints. The system configuration, in this paper is determined using the GA.

Step 3 – the proposed number of switches and their locations then become the NO line sections found by the algorithm. These switches represent the optimum locations that will minimize the objectives for the given system, for the given study period.

Step 1 saves processing time by simplifying the system using elimination of laterals from the system to be studied. It prepares the problem for step 2 by rebuilding the model of the study system with switches only. Laterals are represented by their total load at the branching node. Step 2 then solves a simplified problem without laterals, rather than the traditional complex reconfiguration problem that include laterals in a separate constraint [18]. This system reduction results in reducing processing time.

Fig. 2 gives an overview of the solution algorithm.

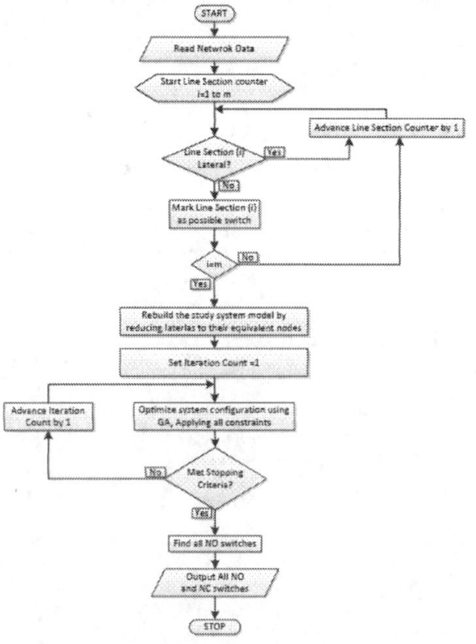

Fig.2. The SA solution Algorithm.

The solution of the optimization problem was obtained using the GA, which is an older algorithm that appeared in the early 1990s [19]. GA (Goldberg 1989) is a search algorithm based on the principle of natural genetics and evolution.

The GA encompasses stopping criteria, which determines when to stop. This includes reaching maximum iterations, obtaining a solution that meets maximum tolerance in comparison to the previous solution, reaching maximum number of population generations, etc.

GA has proven to be a useful approach to address a wide a variety of optimization problems. Being a population-based approach, GA is well suited to solve the multi-objective optimization problems.

GA detailed explanation could be found in many references and therefore it will not be detailed further in this paper.

5 Case study

Several test cases were used to test the proposed model and solution algorithm. The 54-bus real system test case is presented here and discussed in this paper.

This test case consists of 54 buses, including 4 substations and 60 line-sections connecting these buses. The system also includes 2 DGs connected to two different buses and 7 laterals. These DGs are simulated using a P-V model with the assumption of infinite reactive power (VAR) abilities. **Fig.** 3 presents the test system with its original configuration.

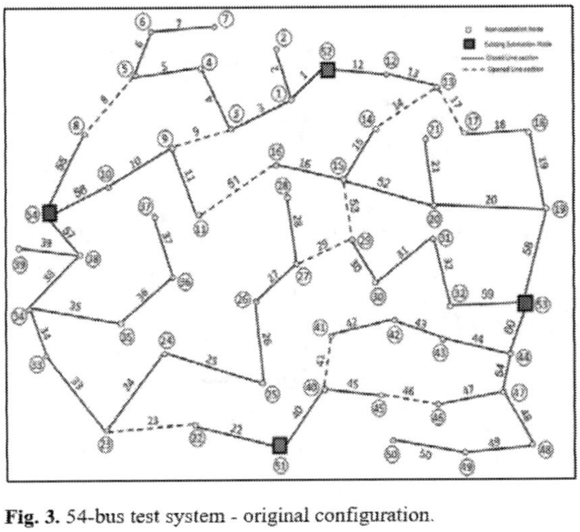

Fig. 3. 54-bus test system - original configuration.

DGs are tied to nodes 4 and 50 with installed capacities of 5 and 10 MW respectively. The total load of the whole system is 4750 MW with an average power factor of 0.95.

Line-sections shown in dashed lines are existing NO sections. The test case result comparison is shown in Table 1. In this test case, the A, B, and C weighting factors stated in equation (4) are taken as unity.

Table 1. Test Case parameters before and after optimization.

	Before optimization (original configuration)	After optimization
URI	26.4748	8.836
SRI	3.4433	3.1382
SVI	5.4	4.2336
AEL (kWh)	601,865.5	436,746.88
Total Objective (F)	162.0273	107.81
NO sections	8, 9, 14, 17, 23, 29, 41, 46, 51, 53	3, 5, 14, 17, 27, 33, 42, 47, 51, 53

Objective function parameters, such as URI, SRI and SVI, are not expected be individually reduced; rather the total objective function value will be reduced. In this example the proposed solution algorithm was also able to reduce each individual parameter. This is usually impacted by the weighting factors, mentioned in equations (4) through (7), given to these parameters. These weighting factors are determined according to the strategy of each utility to identify the priority parameters driving investments.

From Table 1 the solution proposes the use of a new set of NO switches. It is also clear that the solution algorithm has avoided placing switches on laterals such as line-sections 6 and 7, as they were eliminated during the solution application.

6 Conclusion

In this paper a new model for the SA problem was proposed. The new model included a combination of three performance indicators replacing the traditional EENS function and combined with the commonly used AEL function as a multi-objective function aggregated under one representation. Eight constraints were used in the solution for the first time. The new proposed model demonstrates its viability to arrive to an optimum solution considering the modern approaches of smart grids including performance. After testing the model using test systems with variable parameters it can be concluded that the model is a practical implementable model that proposes a solution suitable for finding a trade-off between performance and losses. The model can be easily applied in utilities and is recommended to be used by planners to help make the best investment decisions

References

1. M. Y. Wu, M. I. Ridzuan and S. Z. Djokic, "Smart grid functionalities for improving reliability of rural electricity networks," in *Mediterranean Conference on Power Generation, Transmission, Distribution and Energy Conversion (MedPower)*, Belgrade, Serbia, (2017)

2. S. Ray, S. Bhattacharjee and A. Bhattacharya, "Optimal Allocation of Remote Control Switches in Radial distribution netwrok for Reliability

Improvement," *Ain Shams Engineering Journal - Elsevier B.V.*, (2016)

3. H. Hashemi-Dezaki, H. Haeri-Khiavi and H. Askarian-Abyaneh, "Reliability optimization of Electrical Distribution Systems using Internal Loops to minimize Energy Not-Supplied (ENS)," *Journal of Applied Reasearch and Technology*, (2014)
4. J. C. Lopez, J. F. Franco and M. J. Rider, "Optimization-Based Switch Allocation to Improve Energy Losses and Service Restoration in Radial Electrical Distribution Systems," *IET Generation, Transmission & Distribution Journals*, **10**, 2792-2801 (2016)
5. I. Lim, T. S. Sidhu, M. S. Choi, S. J. Lee and B. N. Ha, "An Optimal Composition and Placement of Automatic Switches in DAS," *IEEE Transactions on Power Delivery*, **28**, 1474-1482 (2013)
6. L. G. W. da Silva, R. A. F. Pereira and J. R. S. Mantovani , "Optimized Allocation of Sectionalizing Switches and Control and Protection Devices for Reliability Indices Improvement in Distribution Systems," in *IEEE/PES Transmision and Distribution Conference and Exposition: Latin America (IEEE Cat. No. 04EX956)*, (2004)
7. L. Zemite, M. Gorobecs, A. Smats, A. Jasevics and A. Levchenkovs, "Genetic Algorithm for Optimization of Power Switch Allocation in Distribution Network," in *IEEE International Conference on Environment, Electrical Engineering, Industrial and Commercial Power Systems, Europe (EEEIC / I&CPS Europe)* , (2017)
8. G. D. Ferreira, D. S. Gazzana, A. S. Bretas, A. L. Bettiol, A. Carniato, L. F. do N. Passos, R. Z. Homma and F. N. Molina, "Multiobjective MILP Model for Optimal Allocation of Automated Switching Devices in Electric Power Distribution Systems," in *Poweer Tech, IEEE Eindhoven Conference*, (2015)
9. "IEEE std. 1366-2012 IEEE Guide for Electric Power Distribution Reliability Indices," IEEE Power & Energy Society, (2013)
10. D. P. Bernardon, M. Sperandio, V. J. Garcia, L. N. Canha, A. D. R. Abaide and E. F. B. DazaG, "AHP decision-making algorithm to allocate remotely controlled switches in distribution networks," *IEEE transactions on Power Delivery*, **26**, 1884-1892 (2011)
11. A. Moradi and M. Fotuhi-Firuzabad, "Optimal switch placement in distribution systems using trinary particle swarmoptimization algorithm," *IEEE Transactions on Power Delivery*, **23**, 271-279 (2008)
12. J. Chen, M. Peng, X. Gao and G. Li, "Multi-objective Distribution Network Planning Considering Invulnerability," in *IEEE 2nd Information Technology, Networking, Electronic and Automation Control Conference (ITNEC)*, Chengdu, China, (2017)
13. P. M. S. Carvalho, L. A. F. M. Ferreira and A. J. C. da Silva, "A decomposition approach to optimal remote controlled switch allocation in distribution systems," *Transactions on Power Delivery*, **20**, 1031-1036 (2005)
14. H. Sindi and E. El-Saadany, "Unified Reliability Index Development for Utility Performance Assessment," *Intelligent Industrial Systems*, **2**, 149–161 (2016)
15. Z. Ghofrani-Jahromi, M. Kazemi and M. Ehsan, "Distribution Switches Upgrade for Loss Reduction and Reliability Improvement," *IEEE Transactions on Power Delivery*, **30**, 684-692 (2015)
16. F. H. Buller and C. A. Woodrow, "Load factor-Equivalent hour values compared," *Electricity World* **92**, 59-60 (1928)
17. R. F. Arritt, R. C. Dugan and T. A. Short, "Determining Loss Factor With the Use of Sequential-Time Simulations," *IEEE Transactions on Industry Applications*, **51**, 1933-1937 (2015)
18. A. Mazza, G. Chicco, A. Russo and E. O. Virjoghe, "Multi-Objective Distribution Network Reconfiguration Based on Pareto Front Ranking," *Springer Science & Business Media, Intelligent Industrial Systems*, Vols. DOI 10.1007/s40903-016-0065-6, (2016)
19. V. Miranda, J. Ranito and L. Proenca, "Genetic algorithms in optimal multistage distribution network planning," *IEEE Transactions on Power Systems*, **9**, 1927–1933 (1994)

Research on Switching Model of Microgrid with Distributed Power Supply

LIU Meiyin[1], *LV* Zhenhua[2], *WU* Beibei[1] , *GUO* Chongyang[1], and *HAN* Pingping[3]

[1] Renewable Energy Research Center, State Key Laboratory of Operation and Control of Renewable Energy & Storage Systems, China Electric Power Research Institute, Beijing, China
[2] Power Grid Technology Center, Power Science Research Institute of State Grid Jiangsu electric power CO,. LTD, Nanjing, China
[3] HeFei University of Technology, Hefei, China

Abstract. The hybrid microgrid with distributed power supply which includes wind turbine (WT), photovoltaic (PV) inverter and power converter system (PCS), not only could be connected to the power distribution network, but also can operate independently. It is realized switch from grid-connected mode to off-grid mode through the microgrid controller. Whether the microgrid can achieve smooth switching from grid-connected to off-grid has great influence on the safety and stability of the power grid. In order to accurately evaluate the influence of the microgrid switch to the distribution network, it is needed to research on the distributed source modeling technology, and establish an accurate microgrid model. In this paper, using a 400V voltage grade microgrid system with PV, WT and PCS as an example， model the internal equipment and the microgrid control system, and establish a microgrid model for simulation. At the meantime, it carries out the microgrid switch testing, and compare with the measured waveform to verify the accuracy of the model. It is confirmed the accuracy of the microgrid model by the contrast result.

1 Introduction

The renewable energy resources are abundant, such as wind power and solar power, but because of their inherent characteristics such as volatility and randomness, in order to improve the system stability, the corresponding energy storage system should be configured to suppress the power fluctuation in practical application [1-3]. The renewable energy resources are used as distributed power supply which are connected to the distribution network in the role of microgrid, and supported each other with the distribution network. It is an effective way to play the functions of the distributed generation system [4]. The integrated utilization of wind power, PV and energy storage is an autonomous system that can realize control, protection and management by itself. It can operate in parallel with the large power grid, and it can also run in isolation [5]. The switching function which switches from grid-connected to off-gird is one of the basic functions of microgrid. In order to obtain accurate microgrid switching model, it is necessary to study the accurate key equipment and control system model of microgrid. Distributed generation and microgrid modeling technology have been widely studied. The calibration method of PV inverter model is proposed in literature [6], but only the accuracy of grid-connected PV inverter model is explained. Literature [7] presents an algorithm for optimal sizing of the WT, PV and the battery energy storage system in the grid-connected configuration of a microgrid, but there is no mention of modeling methods. Literature [8] describes two PV modeling methods and their performance, but it's no analysis of other distributed power access to the microgrid. Literature [9] describes and evaluates the feasibility of control strategies to be adopted for the operation of a microgrid when it becomes isolated. It presents an evaluation of the need of storage devices and load shedding strategies, but it does not analyze the switching capability of microgrid. At home and abroad, researches on microgrid including wind power, PV and PCS have been carried out. However, there is a lack of comparative research on switching strategy simulation and field testing.

Based on this, this paper makes a complete modeling and experimental analysis of the microgrid system composed of WT, PV and PCS. It studies on the dynamic behavior of the microgrid under the grid-connected conditions and the island conditions, and the switching situation between grid-connected and island condition. Taking the microgrid system of the academy of electrical engineering as an example, the internal equipment and control system of the microgrid are modeled and analyzed. The system includes 30kW horizontal axis WT, 5kW rotor WT, 5kW H airfoil WT, 30kW fixed polycrystalline PV power inverter, 75kWh lithium battery energy storage, simulated load and so on.

2 Control strategy

2.1 Microgrid control mode

At present, there are two kinds of control strategies for microgrid: master-slave control and peer to peer [10]. In this paper, the master-slave control method is adopted in the microgrid system. When the microgrid is gird-connected, the battery and other distributed power are controlled by constant power (P/Q control). When the microgrid is in an isolated island operation mode, the bi-directional PCS adopts constant voltage and constant frequency control (U/f control).

The control mode of the microgrid is mainly realized by the switching mode of bidirectional PCS. When the microgrid is connected to gird, the PCS uses PQ control mode to output active power. When the microgrid is under isolated network operation, the PCS maintains the voltage and frequency of the microgrid, and tracks the output power and load fluctuation of the WTs and the PVs.

2.2 Control strategy of PV inverter

The PV inverter uses SPWM modulated voltage and current closed loop control to realize the inverter power decoupling control. Based on the equivalent power transformation, the three-phase stationary coordinate system ABC is converted to the two-phase rotating coordinate d/q system. The external loop voltage, reactive power control and inner loop current control system are established in the d/q coordinate system. The control strategy model is shown in Figure 1. U_{dc} and Q are the measured values of direct voltage and reactive power respectively. U_{dcref} and Q_{ref} are the given reference values of direct voltage and reactive power respectively. i_d and i_q are the measured values of d/q axis component of the AC side current respectively. i_{dref} and i_{qref} are the given reference values of d/q axis component of the AC side current respectively. ia、ib、ic and ua、ub、uc are the three-phase instantaneous current and voltage data. P_{mr} and P_{mi} are the drive sign of inverter.

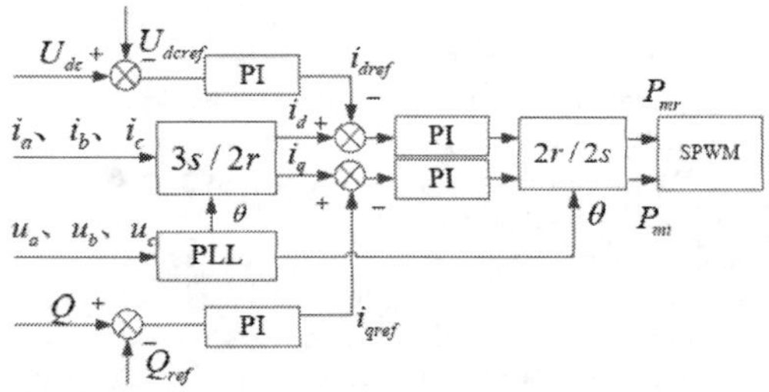

Fig. 1. Schematic diagram of control strategy for PV Inverter.

2.3 Control strategy of direct-driven wind turbine generators

As shown in the Figure 2, the topology diagram of the direct-drive synchronous wind power generation system, the model structure of the direct-drive wind turbine can be divided into several main parts: mechanical control model, direct-drive synchronous motor, generator-side converter model, grid-side converter model and DC link.

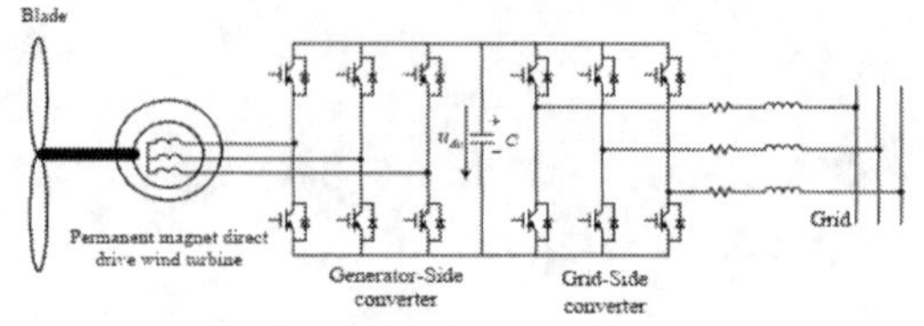

Fig. 2. Topological structure diagram of direct-drive synchronous wind power generation system.

The mechanical control model includes the aerodynamic model of the wind turbine, the control of the pitch angle and the model of the actuator, the shaft model and the MPPT model. The output of the mechanical torque and the maximum power reference value are used as the input of the next module.

The three-phase voltage type PWM converter on the generator side can rectify the alternating current to direct current in amplitude and frequency, and by adjusting the d axis and the q axis components of the stator current of the generator side, the decoupling control of the electromagnetic torque and the stator reactive power (usually set as 0) can be realized, so as to realize the unit power factor of the generator. The factor output also realizes the adjustment of the speed of the wind turbine when the wind speed changes. It can not only make the generator run in the variable speed constant frequency state, but also realize the maximum wind energy capture function under the rated wind speed. The converter on the generator side uses the vector control of the rotor flux orientation of the generator. The control model of the generator-side converter is shown in Figure 3. P_e and P_{eref} are active power measured values and reference values in the generator side. Q_e and Q_{eref} are reactive power measured values and reference values in the generator side. i_{sd} and i_{sdref} are the d-axis current measured values and reference values in the generator side. i_{sq} and i_{sqref} are the q-axis current measured values and reference values in the generator side. $\omega_r L_d i_{sd}/\omega_r L_q i_{sq}$ are d/q axis impedance voltage. U_{sd}/U_{sq} are d/q axis voltage values in the generator side.

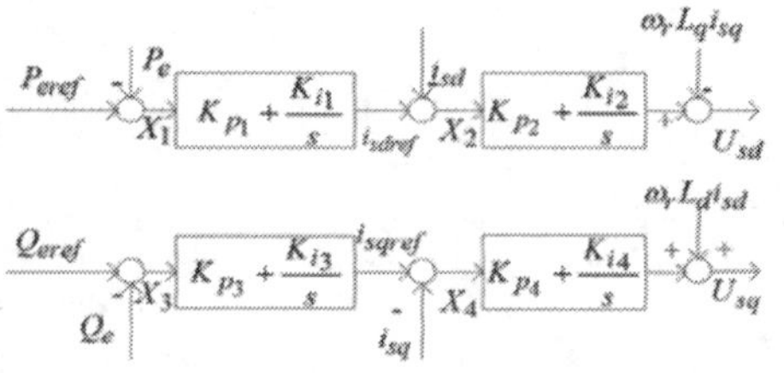

Fig. 3. Generator-side converter and its control model.

The three-phase voltage type PWM converter is used in the grid-side converter, and the three-phase voltage type PWM converter of the grid side controls the reactive current through the PI controller to control the reactive power exchange of the converter and the power grid and regulates the active current to control the constant voltage of the DC bus and the output power of the generator side. Grid side converter adopts grid voltage oriented vector control strategy to realize decoupling control. The control model of the generator-side converter is shown in Figure 4. U_{dc} and U_{dcref} are direct voltage measured values and reference values in the grid side. Q_g and Q_{gref} are reactive power measured values and reference values in the grid side. i_{gd} and i_{gdref} are the d-axis current measured values and reference values in the grid side. i_{gq} and i_{gqref} are the q-axis current measured values and reference values in the grid side. $\omega_r L_g i_{gd}/\omega_r L_g i_{gq}$ are d/q-axis impedance voltage. U_{gd}/U_{gq} are d/q axis voltage values in the grid side.

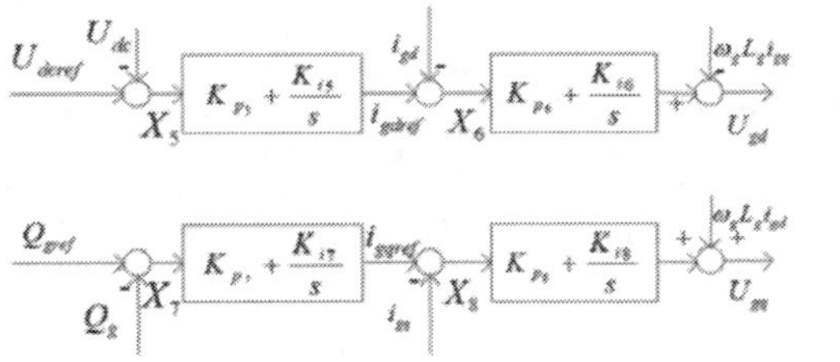

Fig. 4. Grid-side converter and its control model.

2.4 Control strategy of bidirectional PCS

Voltage converter is usually used in bidirectional PCS. An important role of PCS in microgrid is to stabilize power fluctuation in microgrid and realize dynamic balance of power in microgrid. Therefore, when PCS is connected to the grid, PQ control is usually adopted. The control target of PQ control is that the active and reactive power of PCS can follow the instruction value. The PQ control essentially controls the PCS as a controlled current source, and between the PCS and the microgrid is actually equivalent to a parallel connection between an AC power source and a voltage source, while the PCS output voltage is automatically clamped to the grid voltage when the grid is connected.

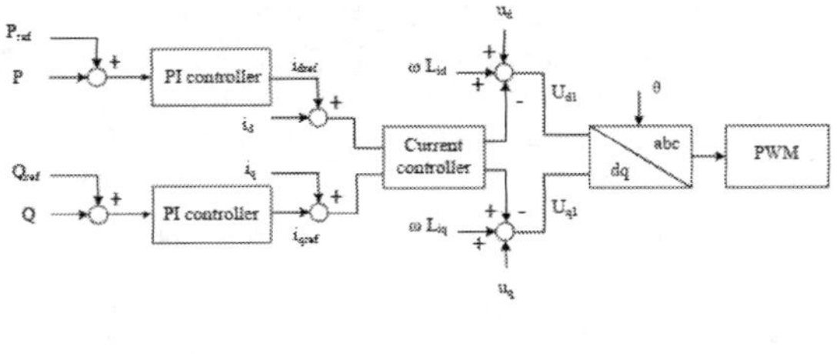

Fig. 5. P/Q control block diagram of PCS.

In the Figure 5, P_{ref} and Q_{ref} are respectively the power given reference values; P and Q are the measured values of the power respectively; id_{ref} and iq_{ref} are the given reference values of the d/q axis components of the AC side current respectively; i_d and i_q are the actual values of the d/q axis component of the AC side current respectively. u_d and u_q are the actual values of the output voltage of the inverter respectively. The reference value of output voltage d/q axis component; L is AC side coupling inductance; θ is voltage initial phase angle.

The AC voltage output of the PCS can be equivalent to a controllable voltage source, and its frequency, phase and amplitude must be coordinated to share the load current. Droop control is used to solve the current sharing problem of multiple PCSs in parallel. And when multiple PCSs run in parallel, each converter can respond to the load changes in a predetermined way. It does not need to obtain data from other converters from the system. Based on local information operation, each converter has the ability to plug and play.

3 Modeling and simulation analysis of microgrid

Based on the above distributed power model structure, building the model in the power system simulation software. The block diagram of PV inverter in DIgSILENT software is shown in Figure 6.

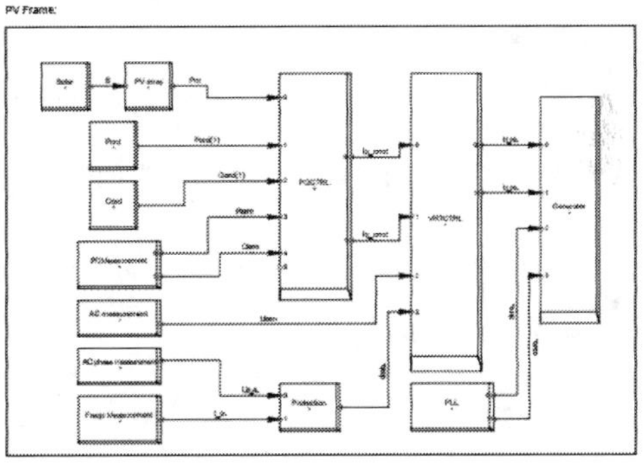

Fig. 6. PV inverter grid-connected control block diagram.

The block diagram of WT in DIgSILENT software is shown in Figure 7.

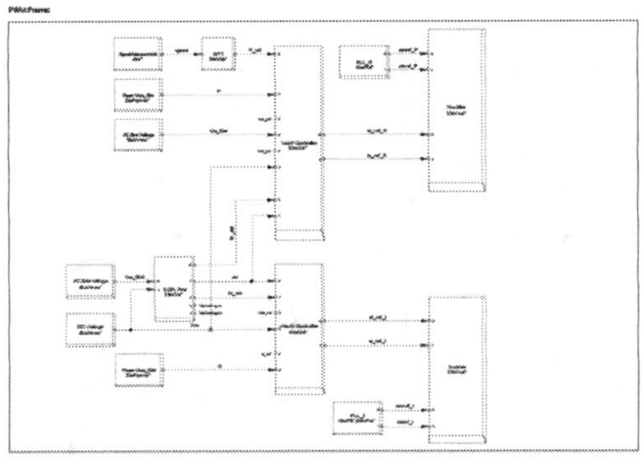

Fig. 7. Direct-driven WT grid-connected control block diagram.

The block diagram of PCS in DIgSILENT software is shown in Figure 8.

https://doi.org/10.1051/e3sconf/20185703008

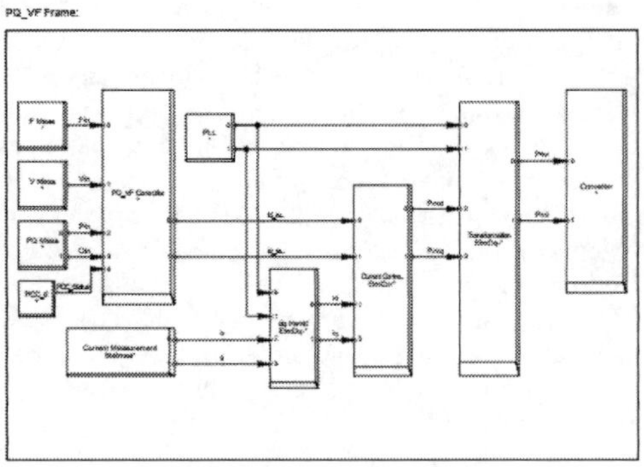

Fig. 8. PCS grid-connected control block diagram.

Take a microgrid as an example, the microgrid includes 30kW horizontal axis WT, 5kW rotor WT, 5kW H airfoil WT, 30kW fixed polycrystalline PV power inverter, 75kWh lithium battery energy storage, wind turbine, PV power inverter, energy storage converter are connected to 0.4kV bus in parallel, after one booster transformer to access 10kV bus, and achieve grid-connected power generate. Based on the DIgSILENT software platform, a model which includes WT, PV inverter and PCS is set up, and the basic network control block diagram is shown in Figure 9. Based on the above model of PV inverter, WT, PCS, and transformer, line and load model are used to simulate the whole microgrid. The power grid model adopts the ideal voltage source, the voltage and frequency of which can be adjusted. It is shown in figure 9.

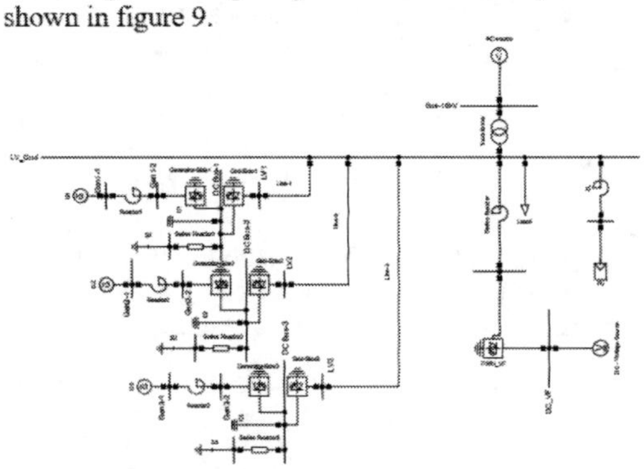

Fig. 9. Microgrid model structure diagram.

The simulation analysis content is to analyze the voltage and frequency characteristics of distribution network and distributed microgrid system in the process of microgrid switching.

Case1: In the grid-connected state, the active power of the WTs is 40kW, the active power of the PV inverter is 30kW, and the microgrid is switched off to the grid at 1.2s.

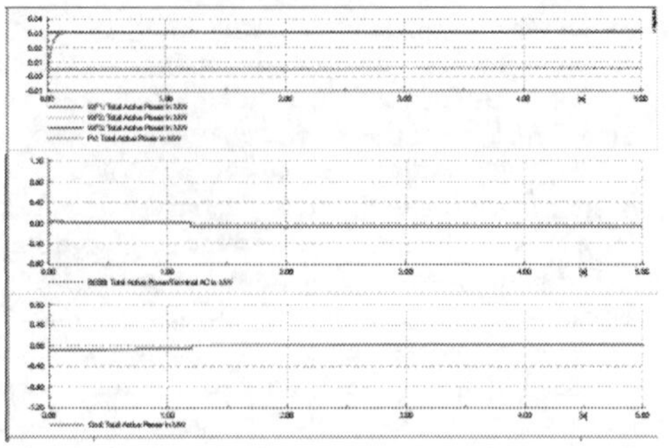

Fig. 10. Active power waveform diagram of PV,WT and PCS in case1.

As shown in Figure 10, the total active power and the PCS (BESS) active power fluctuate when the switch offs at 1.2s. The microgrid is under the island mode after 1.2s.At the time, the PCS supplies the power to the PV and WFs.

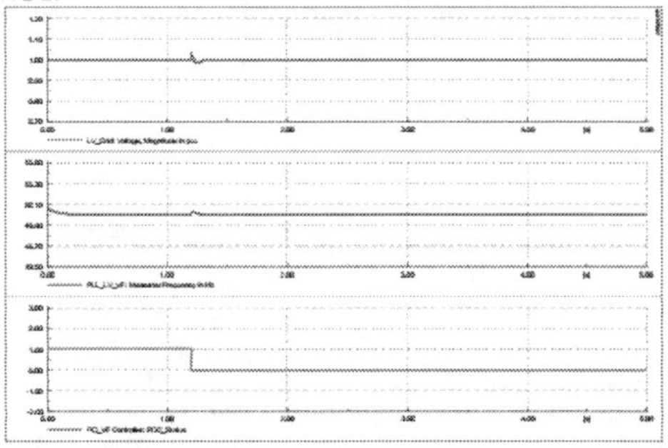

Fig. 11. Voltage and frequency waveform diagram of microgrid in case1.

It is shown that the output voltage and frequency of microgrid are stable when the microgrid is switching in Figure 11.

Case2: In the off-grid state, the load is supplied by the PV, WTs and PCS. The frequency of PCS is set to 50.1Hz, the active power of the WTs is set to 40kW, the active power of the PV inverter is set to 30kW, the load is set to 70kW, and the microgrid is connected to the grid at 1.2s.

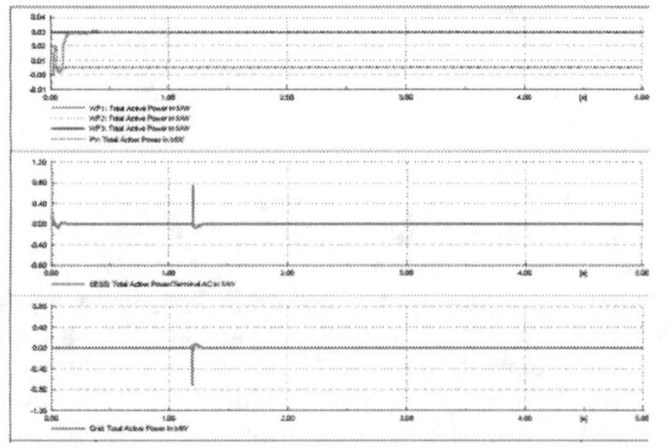

https://doi.org/10.1051/e3sconf/20185703008

Fig. 12. Active power waveform diagram of PV,WT and PCS in case2.

As shown in Figure 12, the total active power and the PCS (BESS) active power fluctuate when the switch on at 1.2s. The microgrid is under the grid-connected mode after 1.2s.At the time, the active power of the PV and WFs supplies to the grid, and the load is connected to the grid too.

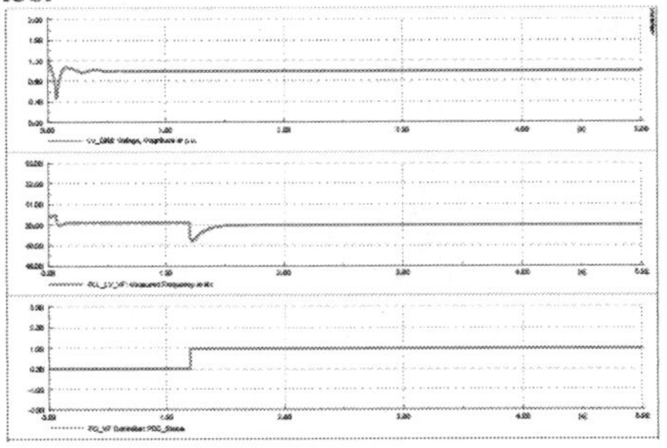

Fig. 13. Voltage and frequency waveform diagram of microgrid in case2.

It can be seen that when the microgrid is switched from the grid to the grid, the frequency of the microgrid fluctuates, but the voltage and frequency can be stably output and can be connected to the grid steadily, which is shown in figure 13.

4 Comparison and verification of measurement and simulation

In order to verify the accuracy of the overall model of the microgrid, microgrid switching tests were carried out in the microgrid system. The test contents are shown in Table 1.

Table 1. Test content.

	Test condition	Test point
On to Off	In the grid-connected state, PV 20kW, PCS 50kW, set be off the grid.	Distributed power supply side 、 point of common coupling (PCC)
Off to On	Case2: In the off-grid state, frequency of PCS is 50.1Hz, PV 20kW, set be connected to the grid.	

Case1: In the grid-connected state, PV 20kW, PCS 50kW, set be off the grid. The test and simulation data contrast diagram is shown in figure 14-15.

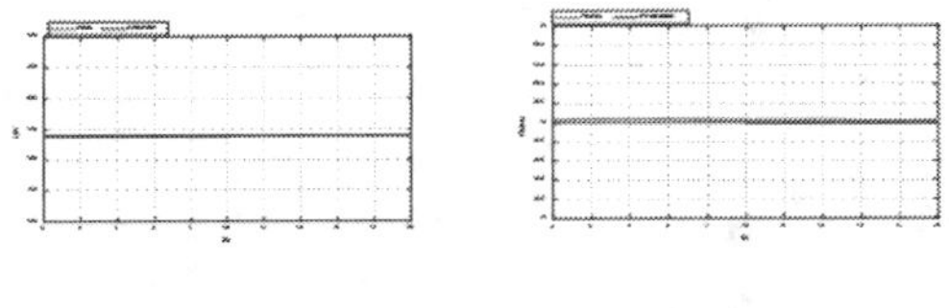

Fig. 14. PCC voltage and frequency contrast diagram in case1.

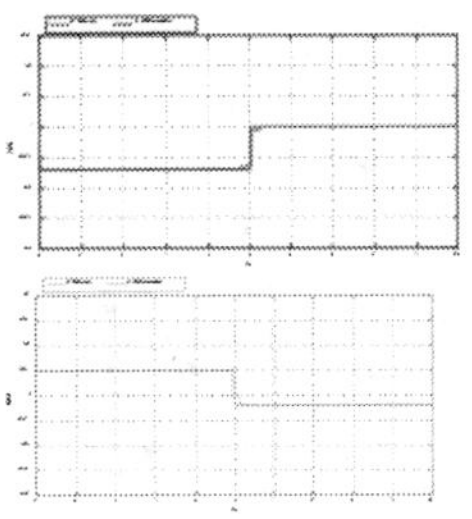

Fig. 15. PCC active power and PCS active power contrast diagram in case1.

In the steady-state interval before and after the switching, and the error calculation in the transient interval of the switching process, three kinds of error values are defined, which are the average deviation F1 of the steady state interval, the average deviation F2 of the transient interval, the maximum deviation F3 of the steady state interval, and the calculation results, as shown in Table 2.

Table 2. Test contentTest error analysis in case 1.

	F1	F2	F3
PCC Voltage	0.0045	0.024	0.015
PCC Frequency	0.0002	0.0006	0.0005
PCC Active power	0.014	0.029	0.062
PCS Active power	0.0024	0.016	0.035

Case2: In the off-grid state, frequency of PCS is 50.1Hz, PV 20kW, set be connected to the grid. The test and simulation data contrast diagram is shown in figure 16-17.

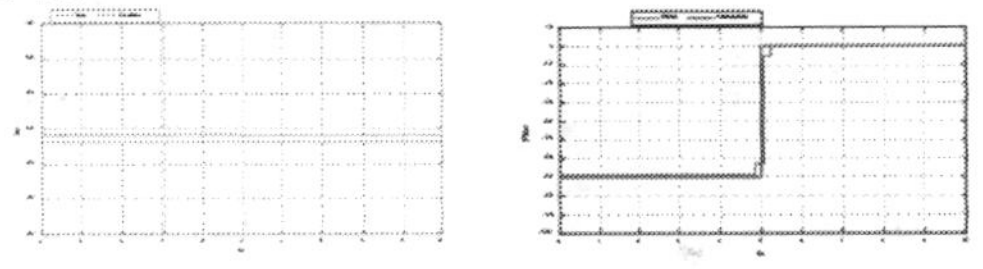

Fig. 16. PCC voltage and frequency contrast diagram in case2.

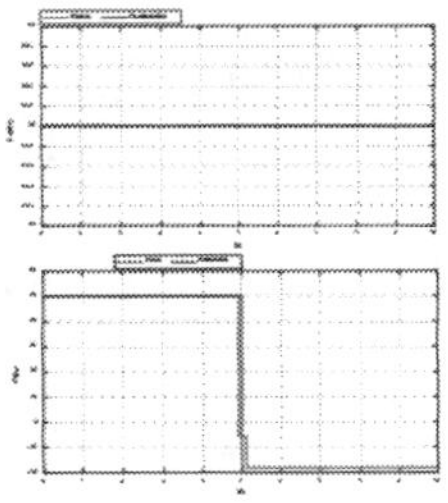

Fig. 17. PCC active power and PCS active power contrast diagram in case2.

It can be seen that the microgrid can run steadily in case 2, and the error between the measured data and the simulation data is small.

Table 3. Test content Test error analysis in case 2.

	F1	F2	F3

PCC Voltage	0.016	0.037	0.033
PCC Frequency	0.0003	0.0006	0.0005
PCC Active power	0.0056	0.029	0.065
PCS Active power	0.0018	0.14	0.16

Through calculation, three kinds of error values which include F1, F2 and F3 are small either, as shown in Table 3. The microgrid model can accurately reflect its running characteristics.

5 Conclusion

This paper takes the 400V AC microgrid system as the object, carries out the simulation and analysis of the measured modeling and Simulation of the 400V microgrid with PV, WT and PCS. Based on the power system simulation software, the key equipment modeling is carried out, and the PV inverter, the WT and the PCS model are combined, and the functions can be realized according to the microgrid. In view of the microgrid's overall parameter testing, a microgrid model is established. Based on the above models, the simulation analysis is carried out for different working conditions, and compared with the measured waveforms for checking and verifying. On the basis of the experimental environment of the Academy of electricity, the field measurement of the microgrid switching is carried out, and the simulation waveform is checked and verified with the measured data. The results prove that the model is accurate and effective.

Acknowledgment

This study is supported by Science and technology project of State Grid Corporation of CHINA "Research on standard framework and certification standard of new energy generation under the background of internationalization" (Grant No. 52110417000F).

References

1. Sundar Venkataraman; Chris Ziesler; Peter Johnson; Stephanie Van Kempen.Integrated Wind, Solar, and Energy Storage: Designing Plants with a Better Generation Profile and Lower Overall Cost. IEEE Power and Energy Magazine, **16** (2018)
2. Jingyang Fang, Xiaoqiang Li, Yi Tang. Energy storage system control strategy to minimize the voltage and frequency fluctuation in the microgird. 2018 IEEE Applied Power Electronics Conference and Exposition (APEC)
3. Huang jingsheng, Liu Meiying, Dongwei, Chen zhilei. Analysis and field-test on Reactive Capability of Photovoltaic plant based on clusters of inverters. Journal of Modern Power Systems and Clean Energy, **5** (2017)
4. YANG Zhangang, WANG Chengshan, CHE Yanbo. A small-scale microgrid system with fexible modes of operation. Automation of Electric Power Systems, **33** (2009)
5. ZHOU Lin, HUANG Yong, GUO Ke, et al. A survey of energy storage technology for micro grid. Power System Protection and Control, **39** (2011)
6. LIU Meiyin, HUANG Jingsheng, ZHANG Junjun, CAO lei. Photovoltaic Grid-connected Inverter Model Validation and Error Analysis based on the Standard of BDEW. Automation of Electric Power Systems, **38** (2014)
7. Umer Akram; Muhammad Khalid;Saifullah Shafiq. Optimal sizing of a wind/solar/battery hybrid grid-connected microgrid system. IET Renewable Power Generation, **2** (2018)
8. Emilien Duverger; Carolina Penin; Philippe Alexandre, et al. PV modeling methods of an off-grid experimental microgrid. IEEE International Conference on Industrial Technology, 2018.
9. J. A. P. Lopes; C. L. Moreira; A. G. Madureira. Defining control strategies for microgrids islanded operation. IEEE Trans on Power Systems, **21** (2006)
10. TIAN Liangyu, TANG Zhong, TIAN Chen, ZHU Ruiting.Research of Microgrid Seamless Switching Based on State Follower. Power System Technology, **41** (2017)

Improved Human Thermal Comfort with Indoor PCM-Enhanced Tiles in Living Spaces in the Arabian Gulf

Albert Al Touma, *Djamel* Ouahrani

Department of Architecture and Urban Planning, College of Engineering, Qatar University, Doha, Qatar

Abstract. Al-Majlis is the living space in residential buildings of the Arabian Gulf, and is where occupants spend most of their time. For this reason, the human thermal comfort in this space is of extreme importance and is often compromised due to hot outdoor weather conditions. In contrast with many thermal discomfort mitigation methods in outdoor environments, which become unadvisable in indoor spaces, this study investigates the effect of adding PCM-enhanced tiles to portions of the indoor envelope on the occupant's thermal comfort and the space cooling energy demand. A simulation model of a space with tight building envelope in Qatar was developed on EnergyPlus with and without the addition of PCM-enhanced tiles. The selected country is a representative location of the Arabian Gulf. Considering different occupant's positions, the addition of the tiles with PCM on their back was found to moderate the mean radiant temperature, operative temperature, Predicted Mean Vote (PMV) and Predicted Percentage of Dissatisfied (PPD), all of which signify an improvement in the human thermal comfort. Lastly, this change in the indoor envelope was found to save 3.3% of the space daily thermal cooling energy demand during one harsh summer representative day.

1 Introduction

Energy consumption has been a hot research topic due to its increasing annual trend, fortunately at a decelerating rate [1]. Among the different energy-consuming sectors, buildings still account for over 40% of the annual energy consumption worldwide [2]. In the Arabian Gulf, this percentage is even higher. In fact, residential and commercial buildings consume around 65% of the total electricity produced in the Kingdom of Saudi Arabia [3] and over 60% of the total energy produced in Qatar [4]. In this region of the world, the living space of the residential buildings, or the so-called Al-Majlis, is of unique importance as it accounts for a significant contributive share.

Al-Majlis is a unique thermal zone in almost each residential building in the Arabian Gulf; it is where occupants spend most of their time, entertain and welcome visitors. Al-Majlis occupants usually sit on the ground level as traditions mandate in the region. Due to its importance, thermal comfort of the occupants in such spaces is crucial and should be critically studied.

As any other living space, Al-Majlis requires a visual sight to the outdoor environment, which is reflected by a Window-to-Wall Ratio (WWR) of at least 30%. As the glazed surfaces are known to be the weakest component of the building envelope, these windows typically cause significant solar transmission and absorption, eventually augmenting the space thermal energy demands and damaging its occupants' thermal comfort [5, 6].

Recently, solutions to thermal discomfort in outdoor environments have been proposed through the use of jackets filled with Phase-Change Materials (PCM). These jackets were found to mitigate the workers' thermal stresses significantly, even in harsh weather conditions [7-9]. However, the applicability of such techniques is troubled in indoor spaces. In other words, the occupants of the living space would not prefer moderating their thermal discomfort through wearing such jackets at home.

Alternatively, techniques applied to the space outdoor envelope have been proposed to decrease its energy consumption and simultaneously enhance the occupants' thermal comfort. These techniques include the design of a solar chimney integrated with passive evaporative cooler and applied on windows [10], the installation of shading devices [11, 12], the adoption of shading and lighting control strategies [13, 14] and the change in the air-distribution system [15]. However, all of these methods are never suitable in all cases; they are inapplicable in humid conditions, sacrifice the visual comfort of the occupants by blocking their sight to outside, and are sometimes extremely expensive to adopt.

Instead of modifying the space outer envelope, changes in the inner envelope have been alternatively suggested. For instance, tiles covering the indoor floor surface were proven to mitigate hot thermal sensations and enhance the occupants' comfort level. To further expand their effect, Phase Change Materials (PCM) was added to these floor tiles. The addition of PCM absorbs the heat coming through the floor during the daytime and releases it at night. Hence, it reduces the peak convective heat load dissipated from the floor and shifts it to a later time of the day. Several studies have been carried out in order to analyze the effect of PCM-enhanced floor tiles on thermal comfort and energy consumption in cold climates [16, 17]. It was found that these tiles do enhance thermal comfort, yet they have a little effect of less than 4% in reducing the heating demand. Nonetheless, little research was found on their application in cooling-dominated regions.

In this study, the effect of adding PCM-enhanced tiles on an energy-efficient Al-Majlis living space during one summer representative day is investigated in terms of thermal comfort and space thermal cooling energy demand in the hot climate of Doha, Qatar. The selected city is a representative location of the Arabian Gulf, as the entire area suffers from harshly hot weather conditions.

2 Methodology

2.1 Al-Majlis living space

Before providing detailed description on the selected Majlis, it is important to point out that the space carried out herein takes into account several energy-efficient measures, which have become popular in Qatar. Hence, any further energy reductions are considered beneficial considering the tight building envelope and selected construction practices.

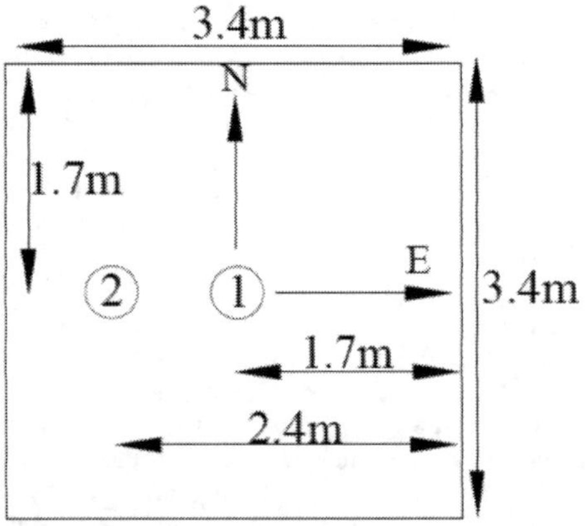

Fig. 1. Schematic representing Al-Majlis dimensions and occupants' positions inside the space.

A typical Majlis living space is picked in this study. As shown in Fig. 1, the space has a floor area of 11.56m^2 and a height of 2.6m [15]. The Majlis is assumed to be exposed to the outside environment from all sides except from the west surface, which is adjacent to other conditioned spaces of the residential building. The space has two windows with dimensions of 2.2m x 1.2m in each of the north and east-oriented walls (window-to-wall ratio of 30%), as these surface directions are known to be hit by relatively less solar radiation intensities throughout the year in comparison with other orientations. The layer-by-layer detailed description of the building envelope, from outside to inside, along with each surface overall heat transfer coefficient (U-value) are reported in Table 1. The selected building envelope has been chosen in according with actual construction practices in the country.

Table 1. Al-majlis thermal envelope selected for this study.

Surface	Specification	U-value (W/m^{2}K)
Wall	2cm Stucco – 20cm Concrete Heavy Weight – 5mm Polyurethane Insulation – 32 mm Gypsum board	2.37
Floor / Ceiling	150mm Concrete Light Weight	3.81
Window	6mm Float Glass – 16mm Clearance – 6mm Float Glass	1.1

Note that the external walls already include a thin insulation layer, which has been more frequently used in Qatar due to the raised awareness for more reduced energy-consuming buildings. Also, note that the window clearance contains 10% air and 90% argon, and the window itself is extremely energy-efficient (SHGC = 0.16) as it has been specifically designed by SAPA for usage in residential buildings in Qatar. Detailed properties of the window are found in [13].

Two representative occupants were picked to analyse their thermal comfort responses. As seen in Fig. 1, the first is located in the middle of the room (i.e. at the position that would be least affected by the addition of tiles as it is the farthest from the wall surfaces) whereas the second is positioned one meter away from the west surface (i.e. closer to a surface with PCM-enhanced tiles).

2.2 Simulation model

2.2.1 EnergyPlus Model

In order to study the thermal performance of the living space and select a thermal comfort model, energy simulation software was deemed necessary. EnergyPlus version 8.5 was selected for this purpose [18]. This software has shown its capability in performing accurate energy analysis using the heat balance method, which is selected by ASHRAE for estimating the buildings' thermal performances [19].

A simulation model of Al-Majlis space was designed on EnergyPlus; all the walls were set to face the outdoor environment with direct sun and wind exposure except for the west wall. This surface was set to *adiabatic* conditions, as recommended by the software documentation for internal surfaces between zones showing little to no great temperature differences [20].

Once the PCM-enhanced tiles are added to the space, they covered the floor/wall as two additional material layers. They were selected to cover the entire floor area and up to a height of 0.6m of the vertical walls surface areas. This distance was chosen because, as per ASHRAE guidelines, a height of 0.6m should be used to □nalyse the thermal comfort of seated occupants [21]. Hence, a new wall was designed and contained the two additional layers while keeping the original wall as is; the first covered the lower part of the space while the latter covered its upper part. The selected PCM was paraffin that has a melting temperature of 30°C [17].

It is of interest to mention that the *ConductionFiniteDifference* heat balance algorithm was used instead of the normal *ConductionTransferFunction* due to its capability in modelling phase change materials, as recommended by EnergyPlus [20]. In addition, the *IdealLoadAirSystem* was used to predict the space thermal cooling energy demands without having to select or design a specific air-conditioning system [20]. The all-day indoor set-point temperature was 24°C. Lastly, the simulation model was set to run for several cycles in order to eliminate the effect of initial conditions.

2.2.2 Thermal comfort model

The thermal comfort of space occupants is of extreme importance and usually depends on several factors, such as the air temperature, humidity and velocity, as well as the space mean radiant temperature, operative temperature and radiant temperature asymmetry [22]. Using these ambient parameters, the Predictive Mean Vote (PMV) is calculated to predict the human thermal sensation. This PMV parameter, and consequently the Predicted Percentage of Dissatisfied (PPD) people, was originally presented by Fanger for the evaluation of thermal environments [23]. Among the different built-in thermal comfort models available on EnergyPlus, Fanger model was selected.

In this model, the human thermal sensation, and consequently its thermal comfort, is related to the body as a whole. In other words, the exact sensation at a specific part of the body is not determined as it requires another segmental body model [10]. This selected Fanger model is applicable in uniform thermal environments where the air ambient parameters (i.e. temperature, humidity, velocity…) do not vary inside the space. In addition, this model is valid in quasi-steady state conditions when the temporal gradient of the parameters affecting thermal comfort do not vary significantly from one time-step to another. Based on the prepared simulation model, all the aforementioned conditions are met.

In order to exclude factors that may affect thermal comfort, the human activity was fixed to a metabolic rate of 1 Met (108W) belonging to seated occupants [21], the clothing was set to *DynamicClothingModelASHRAE55* which fixes the clothing insulation for one day depending on the outdoor conditions [20], and the air velocity inside the space was set to 0.1m/s. Note that the air velocity has a significant effect on the human thermal sensation as moving air, up to a certain point, enhances comfort even at higher space temperatures. For this reason, a conservative value was selected. The rest of the parameter are those affected by the addition of tiles to the space inner envelope and are analyzed through the space mean radiant temperature, operative temperature, PMV and PPD.

In the developed simulation model, the Mean Radiant Temperature (MRT) is calculated using the *anglefactor* method; the view factor between the occupant and each surface, which are calculated by the authors according to ASHRAE [21], are inputted one by one. In comparison with other built-in methods, this method is more accurate as it uses the exact view factor between the occupant and each surface, and weighs their surface temperatures based on that factor according to Equation 1 [20].

$$MRT = \sum F_{Occupant-Surface} \times T_{Surface} \qquad (1)$$

Similarly, the Operative Temperature (OT) is another indicator of thermal enhancement that combines the MRT with the zone temperature, and is analyzed as well in this study. For air velocities of equal or less than 0.1m/s, the operative temperature can be simplified and calculated according to Equation 2:

$$OT = \frac{T_{Zone} + MRT}{2} \qquad (2)$$

According to Fanger, the PMV is a subjective thermal sensation that ranges between -3 and 3 representing very cold and hot sensations, respectively. This parameter is used to assess the human thermal response in comparison with the ideal value of zero that represents thermal neutrality. Hence, the closer the PMV is to the range of -0.5 to 0.5, the better the thermal comfort is. However, since the PMV is a subjective response, it is often combined with the PPD that estimates the percentage of people dissatisfied by the predetermined thermal sensation. Thus, these four parameters are selected to analyze the enhancement in thermal comfort due to the addition of PCM-enhanced tiles. Lastly, the space cooling thermal energy demand is investigated before and after the addition of the tiles to look for potential energy savings.

2.3 Summer representative day

In order to analyze the hourly response of the developed simulation model in hot weather conditions, one representative summer day was picked. July 21st was selected where the hourly temperature, relative humidity and beam and diffuse solar radiations data were extracted from the Doha weather file recorded for the year of 2015 [24]. The outdoor temperature during this day ranged between 32.8°C and 43.0°C whereas the outdoor relative humidity varied between 9% and 67%. On the other hand, the direct and diffuse solar radiations peaked to 841W/m^2 and 182W/m^2, respectively. In the developed model, one *SummerDesignDay* option was selected for the simulation where the variations in the aforementioned outdoor weather parameters were inputted as hourly profiles.

3 Results and discussion

Before going deeper in the results analysis, it is of interest to mention that, analytically speaking, the addition of PCM-enhanced tiles on portions of the walls splits each of them into two. According to Equation 1, the number of vertical walls is doubled and the new view

https://doi.org/10.1051/e3sconf/20185704001

factors between the occupant and each of them is recalculated. More importantly, the view factor between the occupant and the lower portion of the walls where tiles are added is significant, despite their smaller surface areas in comparison with the upper original walls.

Since the occupant activity and clothing as well as the air velocity inside the space were fixed, any changes in the human thermal comfort would be caused solely by the addition of the PCM-enhanced tiles. These tiles will have an inner space surface temperature that is different from those of the walls and, due to angle factor between them and the occupant, will unarguably affect his thermal response. The hourly mean radiant temperature (MRT), operative temperature (OT), PMV and PPD with and without the addition of PCM-enhanced tiles for both occupants are plotted in Figs. 2 and 3.

3.1 Occupant 1

In the base-case (i.e. before the installation of tiles), the MRT ranged between 27.6°C at 5 hours and 34.7°C at 15 hours. This trend shrank to range between 28.3°C and 34.0°C when the tiles were added, as shown in Fig. 2(a). In other words, the addition of the tiles reduced the MRT in the times between 6 hours and 21 hours and increased it slightly during the rest of the day. As anticipated, the addition of tiles with PCM on their back reduces the inner surface temperatures of the walls, eventually dropping the MRT. Consequently, the convective heat load from these surfaces is reduced as heat gain is being absorbed by the PCM. However, during the early and later times of the day, the PCM releases the pre-absorbed heat to the indoor space, thus increasing the tile surface temperature and eventually the MRT.

Based on the definition of the operative temperature represented in Equation 2, and considering that the zone air temperature is fixed, the changes in MRT affected the OT analogously. Yet, the changes in OT are smoother; it varies between 25.7°C and 29.4°C before the addition of tiles and between 26.1°C and 29.0°C after their addition (refer to Fig. 2(b)).

As seen in Fig. 2(c), the changes in PMV were analogous to those of the MRT and OT. Before the addition of tiles, PMV increased during the daytime, reached a value of 1.0 at 15 hours and decreased back down for the rest of the day. Recall that a PMV of zero represents thermal neutrality and a value greater than 0.5 represents thermal discomfort caused by hot sensations. In this case, this condition was always violated during the hours between 9 and 20 hours. This led a dissatisfaction percentage as high as 26.2%, as shown in Fig. 2(d). However, since the addition of the tiles decreased the MRT and OT, it reduced the PMV during most of the hours of the day. Although its peak value is 0.9, the PMV values were moderated throughout the day and became always closer the PMV recommended limit. This moderation led to a decline in the percentage of dissatisfaction whose peak value went down to 21.6%.

Note that an improvement in thermal response is noticed during most of the hours when the living space is expected to be occupied. In early hours of the morning and late hours of the day, during which the space is not expected to be occupied, PMV values increased, yet never violated the 0.5 limit.

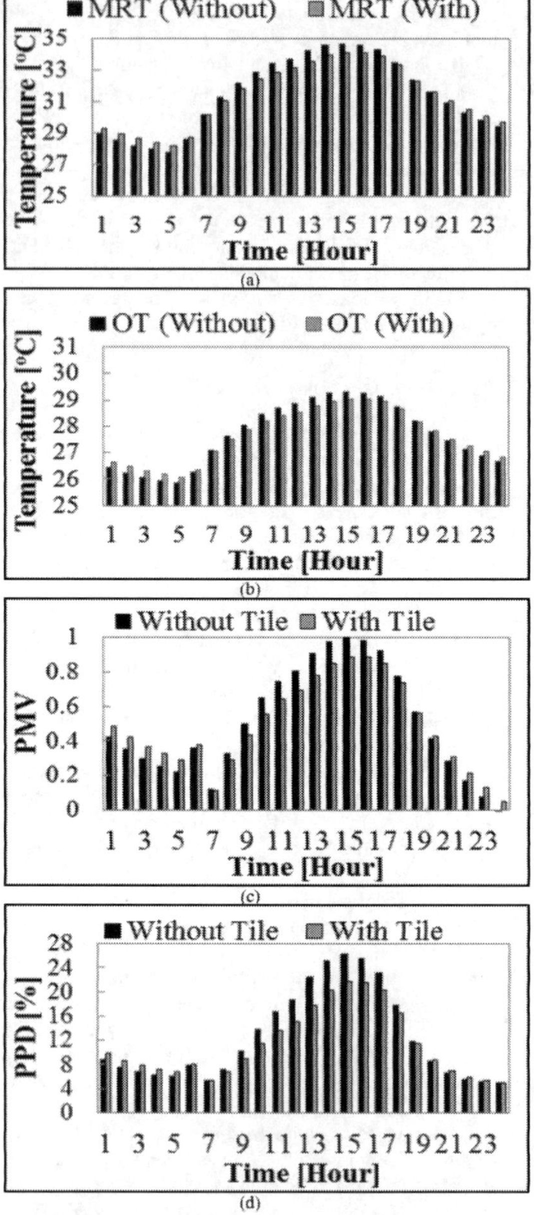

Fig. 2. Hourly results of (a) Mean Radiant Temperature (MRT), (b) Operative Temperature (OT), (c) Predicted Mean Vote (PMV) and Predicted Percentage of Dissatisfied (PPD) for Occupants 1 and 2.

3.2 Occupant 2

The results pertinent to the second occupant are comparable to the first one. Yet, since this occupant is seated closer to the west surface with PCM-enhanced tiles (i.e. farther away from the north and east windows), the benefits of the changes in the indoor envelope are amplified. For instance, and in contrast with the first occupant, the MRT and OT at the second occupant's location decreased throughout the entire period with almost no counter-effects during the late hours of the day, as seen in Figs. 3(a)-(b).

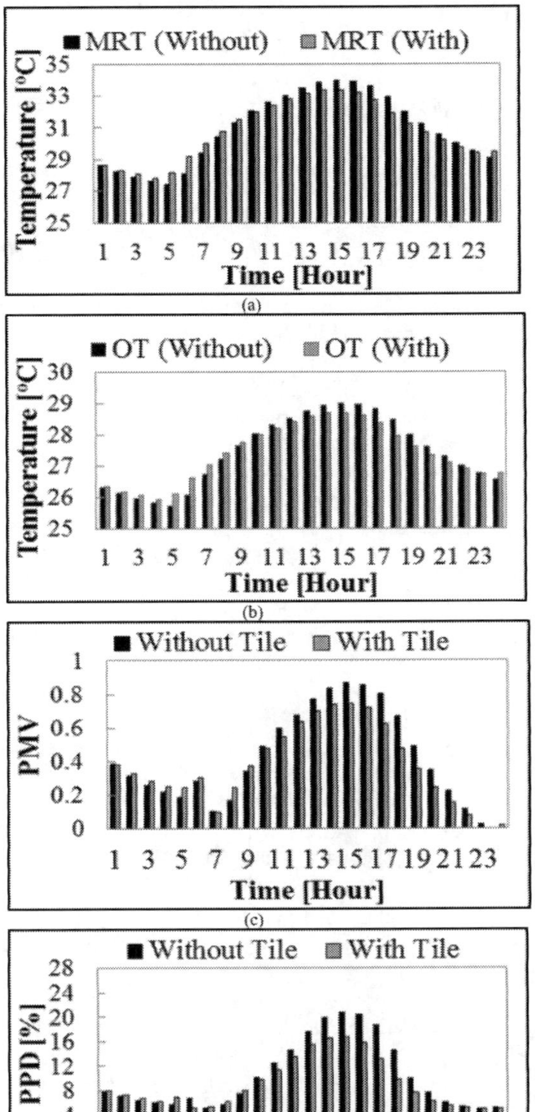

Fig. 3. Hourly results of (a) Mean Radiant Temperature (MRT), (b) Operative Temperature (OT), (c) Predicted Mean Vote (PMV) and Predicted Percentage of Dissatisfied (PPD) for Occupants 1 and 2.

Consequently, these results affected the PMV and PPD analogously. Due to this occupant's larger angle factor with the west surface, which is more moderate than the rest of the space surfaces, his PMV trend shifted down to become closer to the 0.5 proposed limit (refer to Fig. 3(c)) and consequently the PPD decreased simultaneously. In fact, the peak PPD decreased from 20.8% to 16.6%, as shown in Fig. 3(d).

Since the position of the second occupant is more realistic in living spaces, the addition of the PCM-enhanced tiles to portions of the space indoor surfaces is analytically justified as it enhances thermal comfort whenever the space is expected to be occupied.

3.3 Energy savings

Lastly, it would be important to mention that the space daily cooling thermal energy demand before and after the installation of PCM-enhanced tiles decreased from 30.6 kWh to 29.6 kWh. Although it is a slight reduction, the addition of these tiles ended up saving 3.3% of the space cooling demand.

4 Conclusion

Due to the importance of human thermal comfort in living spaces, or the so-called Al-Majlis in the Arabian Gulf, the changes in occupant thermal response and space cooling energy demand caused by the addition PCM-enhanced tiles was investigated. These tiles were selected to cover the entire floor area as well as portions of the vertical walls surface area from the ground level up to a height of 0.6 meters. This distance was chosen as it represents the typical height of a seated person as per ASHRHAE guidelines.

A simulation model of a Majlis space was developed; it considers a tight building envelope containing insulations layers and efficient glazing systems, and two occupants seated differently inside the space. Results have shown that the addition of PCM-enhanced tiles was able to moderate the mean radiant temperature and operative temperature. Consequently, the PMV and PPD trends shrank. During the hour times when the living space is expected to be occupied, the addition of tiles shifted the PMV closer to the comfort range and was found to save 3.3% of the space daily cooling energy demand during one harshly hot summer day in Doha, Qatar. Should the same analysis be conducted on another, less tight living space, energy savings would be intensified. Similarly, the improvement in human thermal comfort is expected to amplify whenever these tiles cover larger portions of the vertical walls or when the original thermal discomfort is greater due to larger glazing surfaces.

Acknowledgment

This publication was made possible by NPRP 7 – 486 –2 – 185 from the Qatar National Research Fund (a member of Qatar Foundation). The findings achieved herein are solely the responsibility of the authors.

References

1. Petroleum, B. BP statistical review of world energy. (2017)
2. Ghiaus C, Inard C. Energy and environmental issues of smart buildings, A Handbook for Intelligent Building, 26–51 (2004)
3. Said, S. A. M., Habib, M. A., & Iqbal, M. O., Database for building energy prediction in Saudi Arabia. Energy conversion and management, **44** (2003)
4. Ayoub, N., Musharavati, F., Pokharel, S., & Gabbar, H. A, Energy consumption and conservation practices in Qatar—A case study of a hotel building. Energy and Buildings, **84** (2014)
5. Al Touma, A., Ghali, K., Ghaddar, N., & Ismail, N., Solar chimney integrated with passive evaporative cooler applied on glazing surfaces. Energy, **115**, (2016)
6. Al Touma, A., & Ouahrani, D. Experimental analysis of double and triple glazed façades with different shading devices in Qatar. In Sustainable and Renewable Energy Engineering (ICSREE), 2017 2nd International Conference. IEEE. 38-41 (2017)
7. Hamdan, H., Ghaddar, N., Ouahrani, D., Ghali, K., & Itani, M. PCM cooling vest for improving thermal comfort in hot environment. International Journal of Thermal Sciences, **102**, 154-167 (2016)
8. Itani, M., Ouahrani, D., Ghaddar, N., Ghali, K., & Chakroun, W. The effect of PCM placement on torso cooling vest for an active human in hot environment. Building and Environment, **107**, 29-42 (2016)
9. Ouahrani, D., Itani, M., Ghaddar, N., Ghali, K., & Khater, B. Experimental study on using PCMs of different melting temperatures in one cooling vest to reduce its weight and improve comfort. Energy and Buildings. (2017)
10. Hweij, W. A., Al Touma, A., Ghali, K., & Ghaddar, N. Evaporatively-cooled window driven by solar chimney to improve energy efficiency and thermal comfort in dry desert climate. Energy and Buildings, **139**, 755-761, (2017)
11. Li, C., Tan, J., Chow, T. T., & Qiu, Z. Experimental and theoretical study on the effect of window films on building energy consumption. Energy and Buildings, **102**, 129-138, (2015)
12. Li, D. H., Lam, T. N., Wong, S. L., & Tsang, E. K. Lighting and cooling energy consumption in an open-plan office using solar film coating. Energy, **33**, 1288-1297 (2008)
13. Al Touma, A., & Ouahrani, D. Shading and day-lighting controls energy savings in offices with fully-Glazed façades in hot climates. Energy and Buildings, **151**, 263-274 (2017)
14. Lee, K. H., Kim, T., Lee, G., & Lee, J. Cooling load reduction effect and its mechanism in between-glass cavity and venetian blind operation during the summer season. In Building Simulation. Springer Berlin Heidelberg. **6**, 351-364 (2013)
15. Hweij, W. A., Ghaddar, N., Ghali, K., & Habchi, C. Optimized performance of displacement ventilation aided with chair fans for comfort and indoor air quality. Energy and Buildings, **127**, 907-919 (2016)
16. Barbour, J. P., & Hittle, D. C. Modeling phase change materials with conduction transfer functions for passive solar applications. Journal of solar energy engineering, **128**, 58-68 (2006)
17. Stamatiadou, M., Antonopoulos, K., & Founti, M.. Energy savings using floor tiles with phase change materials in a S. European country. In Proceedings of 11th international conference on thermal energy storage; Effstock. (2009)
18. U.S. Department of Energy, EnergyPlus Energy Simulation Software. http://apps1.eere.energy.gov/buildings/energyplus. (2013)
19. ASHRAE, Chapter 19 energy estimating and modeling methods, in: ASHRAE Fundamentals Handbook, American Society of Heating, Refrigerating and Air-Conditioning Engineers., Atlanta, USA, (2009)
20. E. Doe, 6.0 Input/Output Reference: The Encyclopedic Reference to EnergyPlus Input and Output, US Department of Energy., (2010)
21. Standard, A. S. H. R. A. E. Standard 55-2010:"Thermal Environmental Conditions for Human Occupancy"; ASHRAE. Atlanta USA. (2010)
22. McQuiston, F. C., & Parker, J. D. Heating, ventilating, and air conditioning: analysis and design. (1982)
23. Fanger, P. O. Thermal comfort. Analysis and applications in environmental engineering. (1970)
24. Online Resource, http://weather.whiteboxtechnologies.com.

A Study on the cooling effects of the park dispersions applied in PUD

Yi-Cheng Chiang [1] and *Hong-Yun* Lin [2]

[1]Assistant professor, Department of Landscape and Urban Design, Chaoyang University of Technology
[2]Department of Landscape and Urban Design, Chaoyang University of Technology

Abstract. Urban heat-island effect causes the vicious cycle of city high temperature. The collocation of green park space can effectively cool down the heat of city. This research studies the cooling effects of the park dispersion by simulations of PUD projects. In this study, 3 different schemes were developed and simulated by CFD software. The results show that the more disperse the parks are, the lower the site temperatures.

Keyword. Urban thermal environment, Cold-island effect, Park scale, Park dispersion

1 Introduction

Most of Taiwan cities have developed according to urban plans and guidelines established in middle 19th century. Nowadays many of them come into a renewal period. Facing the climate of Urban Heat Island(UHI), climate responsive urban design requires comprehensive epistemology in different fields. Green park is an important element of urban composition. Many studies have found that the allocation of green park can effectively reduce the urban high temperature, which is called Cold Island Effect(CIE), and different scales of green park can produce different effects of CIE. It could be profound method for climate responsive urban design to suggest a green park configuration type with more effective CIE. This study investigates on the cooling effects of the park dispersion in order to find out the better dispersion types for climate responsive urban planning and design. The main issue is that in the limitation of green park area, how will various park dispersions impact the urban temperature?

2 Literature Review

The study reviews three kinds of papers, including Cold-island effects of parks, Thermal environments of different park allocations, Functions of different park scales, and Urban climate scale.

2.1. Cold-island effects of parks

The park green space has the function that mediates the climate, alleviates the hot environment in the city, this phenomenon is called the urban cool-island. Urban park green space compared to other urban areas have a lower temperature, the main reason for the park has a higher green cover area, the park's plants can produce shading and evaporation effect, can avoid direct sunlight surface and artificial facilities, endothermic caused by temperature rise, and through the plant body water distribution, increase the humidity in the environment away from heat.

2.2. Thermal environments of different park allocations

Many studies found that the park green space can effectively reduce urban high temperature, which in the park spatial distribution of environmental cooling effect of the study, C. T. Lin [1][1] found that the more dispersed park space distribution, the better the effect. In addition, the study on the effect of cooling in size and scale parks, M. H Lee [2][2] found that the range of cooling caused by dispersing several small parks is not significantly different from that of the whole big park. Wenqi Lina,b [3][3] found that the cooling range of the park in different directions, the maximum cooling range reached 840 M, and the minimum cooling range of only 35 M. And regardless of the size of green space, the minimum temperature range is generally below 100 M, the median is generally below 200 M. This study refers to these three articles, further discussion of the decentralized model of the park, the scheme simulation for the cooling efficiency of this model.

2.3. Functions of different park scales

According to Taiwan CPAMI and the Taizhong government definition, the park green space function has the leisure, the landscape, the conservation and the Urban disaster prevention, provides the ecology, the society, the environmental benefit. The neighbourhood park in Taiwan's urban Park, 0.5-4 hectares, is the smallest unit of leisure and social gatherings [4][4]. Therefore, this study defines the neighbourhood Park below 4 hectares as a small-scale park, including Neighbourhood Park, Pocket Park. Compared with large-scale parks, small-scale parks are scattered in high-density cities. It has the characteristics of flexible location, small area and dispersed distribution, and it can in the city corners and has the most influence.

2.4. Urban climate scale

Ground environment will have an impact on atmospheric air layer, the climate of the ground layer is medium or small-scale climate, when the large-scale climate pattern performance is obvious, the microclimate scale performance is not obvious, conversely, the performance is not obvious, will be dominated by the small scale climate the lower atmosphere. (Fig. 1). In different scales, the degree of urban climate change and the impact factors are different, Masatoshi Yoshino [5][5]. According to different regions, the climate is divided into different scales. (Table 1) Oke [6][6] urbanization and the influence of human activities will form local climate, according to different scales of the climate characteristics of the formation of the four regions: hints Bound Layer, hints canopy, hints Plume and Rural boundary Layer. (Fig. 2). The urban boundary layer hints Bound layer are mainly affected by the flow of heat energy from urban air pollutants and structures. The study explores the climate of the urban boundary layer, which is defined by the Man Mino microclimate, the horizontal range 10m-10km, and the vertical range 10cm~1km. (Table 1)

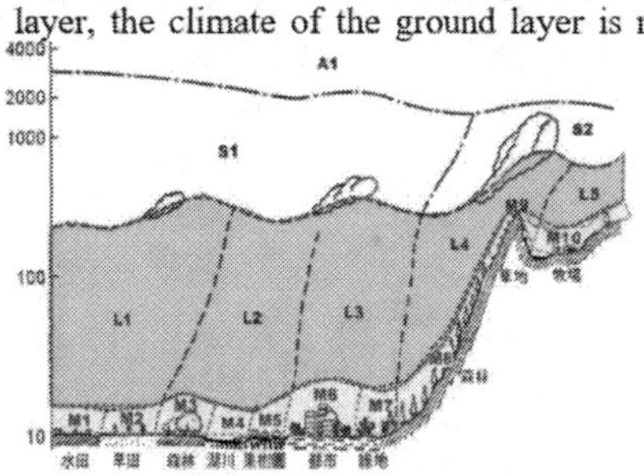

Figure 1. Climate scale diagram

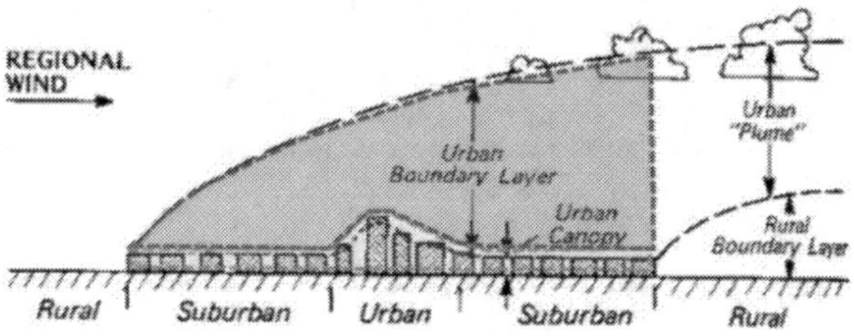

Figure 2. The urban plume

Table 1. Climate scale diagram

Climate	Block number	Horizontal range (m)	Vertical range (m)	Climate phenomenon	Climate phenomenon time
Micro-climate	M1~M10	10^{-2}~10^{2}	10-2~2×100	Paddy climate, Greenhouse climate	10-1~2×101 sec.
Micro-climate	L1~L5	10^{1}~10^{4}	10-1~103	Frost Road, Beveled warm zone	101~104 sec.
Meso-climate	S1~S2	10^{3}~2×10^{5}	100~6×108	Urban climate, Basin climate	103~105 sec.
Marco-climate	A1	$(2\sim4)\times10^{5}$~10^{7}	100~2×105	Climate zone, Seasonal wind band	105~107 sec.

Based on the literature review, we know that the specific park distribution is beneficial to the cold island effect, but the literature in the past has been to simulate the existing environment, without taking into account the actual space structure in the physical environment and the configuration of the building, it is not known how to apply the space of urban development in the future, this study considers the applicability of the future, the paper will make further discussion on the effect of the dispersed mode of green park in urban planned development unit.

3 Method

By the CFD software Phoenics, the research method uses the scenarios simulation on PUD (Planned Unit Development) 12 of Taichung City. All the scenarios are assumed to be on the same district street networks and buildings codes based on the PUD original guidelines and masterplan. The impact of buildings and streets on microclimate are excluded. The total areas of green parks in each scenarios are controlled no more than the given area in the original PUD masterplan.

3.1. Experimental design

3.1.1 Experimental framework

In the limitation of total green park area, how will various park dispersions impact the urban temperature? Since the study assumes that the dispersion types of park will affect the cooling effect, the dependent variable is the atmospheric temperature and independent variable is park green space distribution type. The study controls variables as Table 2. The procedure of experimental framework is illustrated as Figure 3.

Table 2. CFD numerical simulation PHOENICS settings

Parameter	Setting value
time	Summer, May, noon
Original temperature (°C)	30.7
Wind speed (m/s)	10
Green coverage rate(%)	100 (Arbor)

3.1.2 Site

The research select the PUD 12 of Taichung City as the study site. Its main land use zoning is residential areas with an area of 44.3044 hectares, accounting for 57.61% of the site area. Within the planning area. Public facilities include a central park with area of 6.1528 hectares, accounting for 8% of the site area, as well as schools, parking areas and square and parking spaces (Fig.4). Even distribution principle.

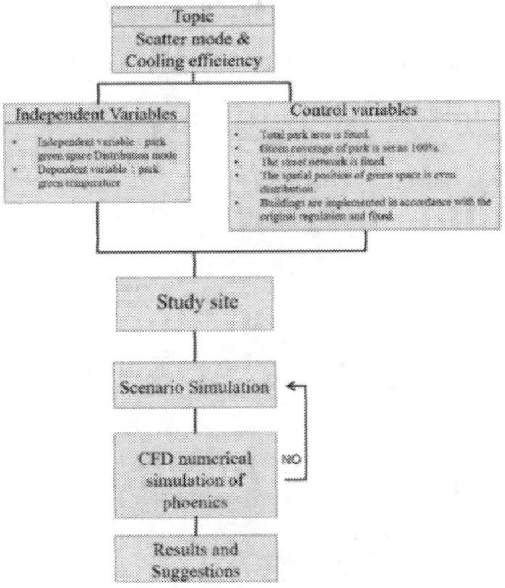

Figure 3. Experimental framework

Figure 4. Study site is the PUD 12 of Taichung City

3.2. Experimental steps

This study scattered the central park area by three dispersion types. Each dispersion type is demonstrated as an individual scenario of PUD. The original project of central park with 6 hectares is designated as the control group. Three dispersion types are the 3-hectare type, the 0.5-hectare as district park type, neighbourhood park type, and the 0.04-hectare type as pocket park. (Fig.5.) All three types are adapted into the PUD street network and adjust their sizes or amounts to fit in blocks. Each of the three types accounts for a lower ratio than the original central park. According to the PUD guidelines, the buildings is three-floors mass, which is demonstrated as 9M high.

E3S Web of Conferences **57**, 04002 (2018) https://doi.org/10.1051/e3sconf/20185704002
ICSREE 2018

• Scenario I (near 3 ha): The whole unit is divided into two sub-areas by the street network. So the 6-hectare central park is dispersed into two parks. Each is located near the center of sub-areas. The total park areas account for 7.08%.

• Scenario II (near 0.5 ha): The whole unit is divided into six sub-areas by the street network. 6-hectare central park is dispersed into 12 pieces of 0.5-hectare parks. These 12 pieces are distributed averagely into six sub-areas. The total park areas account for 7.12%.

• Scenario III(0.04 ha): 6-hectare central park is dispersed into 140 pieces of 0.04-hectare parks. These 140 pieces are distributed averagely around the whole site. The total park areas account for 7.43%.

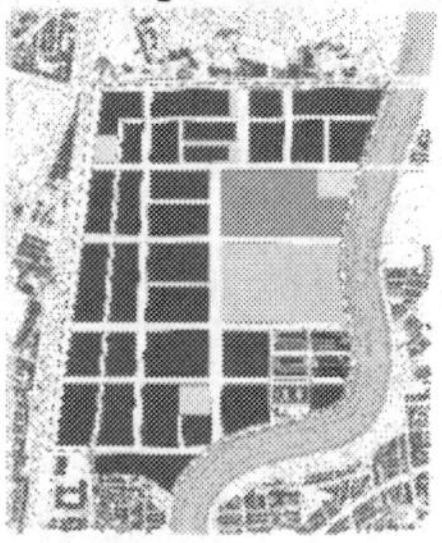

Figure 5.1. Original plan: 6 ha.

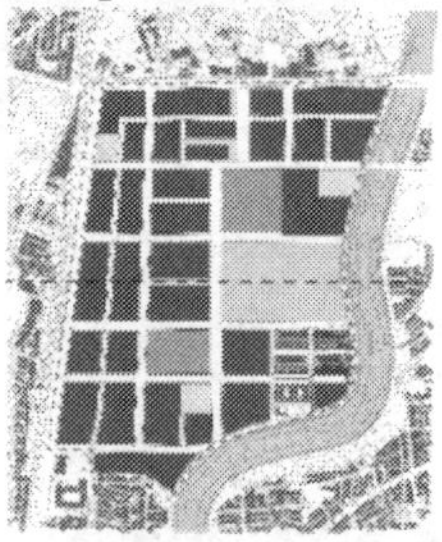

Figure 5.2. Scenarios I (near 3 ha.)

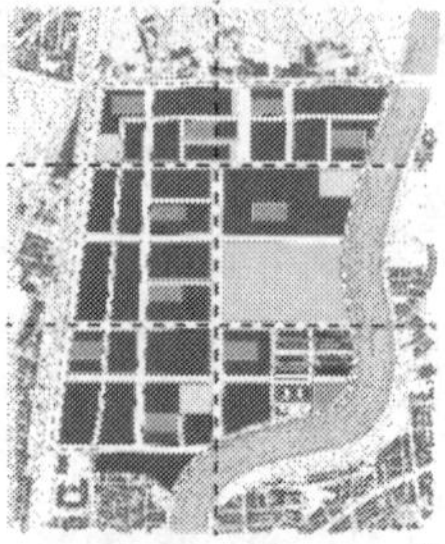

Figure 5.3. Scenarios II (0.5 ha.)

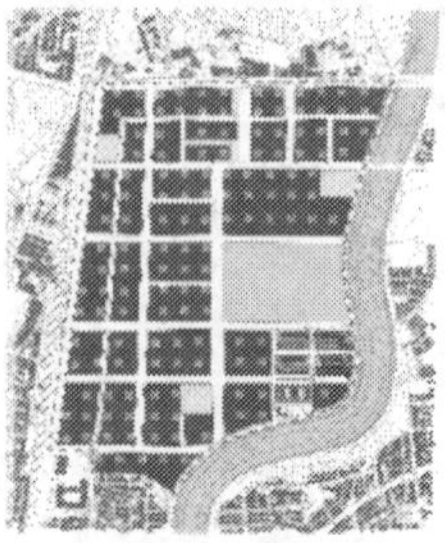

Figure 5.4. Scenarios III (0.04 ha.)

4 Results

In order to discover the temperature patterns in different heights, all four scenarios are simulated in five height, including 5M, 9M,which are among buildings, 10M, 13M, and 16M, which are above buildings. (Fig.6) All the simulations are shown as Fig.8.1~ Fig.8.5. The temperature legend is coloured as Fig.7. The results are summarized as follows.

There is a climb-up point of temperature line in each scenario at 10M height. It is a logical phenomenon because the roofs of buildings would collect heat from sun and radiate the heat, which causes the temperature rise. However, on analysis of temperature pattern, the Scenario III(0.04 ha) always has the most cooler area(shown as light blue to deep blue), and the secondary is Scenario II(near 0.5 ha), the third is Scenario I (near 3 ha), and the original plan has the fewest cooler area. Similarly, on analysis of average temperature (Fig.9), the line of Scenario III (0.04 ha) always runs at the lowest, which means the coolest. The bottom-up sequences of other lines are Scenario II, Scenario I, and the original plan. The results reveal that the more dispersed the park, the lower the average temperature, which means a better cooling effect. Furthermore, the differences of temperature between scenarios are bigger in the height from 9M to 13M, and tend to convert at higher altitude. It reveals that the difference of cooling impact is limited in an extent, which is from the top above of the trees to about seven meters higher. All these conclusions are suggested to the climate responsive urban planning and design.

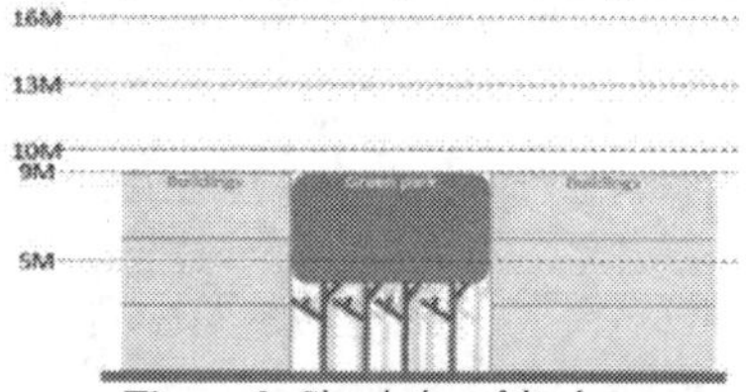

Figure 6. Simulation altitudes

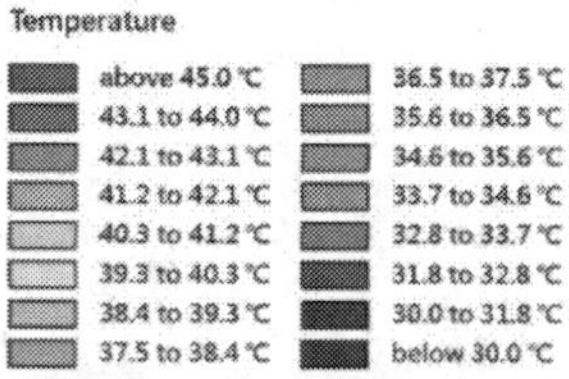

Figure 7. Temperature legend

Figure 8.1.1. Hight_5M
Original plan: 6 ha.

Figure 8.1.2. Hight_5M
Scenario I: 3 ha

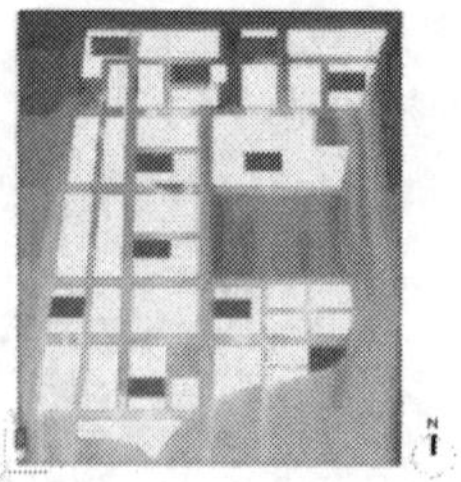

Figure 8.1.3. Hight_5M
Scenario II: 0.5 ha

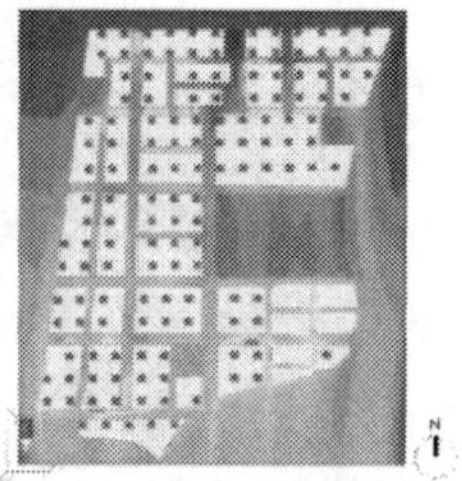

Figure 8.1.4. Hight_5M
Scenario III: 0.04 ha

https://doi.org/10.1051/e3sconf/20185704002

Figure 8.2.1 Hight_9M
Original plan: 6 ha.

Figure 8.2.2 Hight_9M
Scenario I: 3 ha

Figure 8.2.3 Hight_9M
Scenario II: 0.5 ha

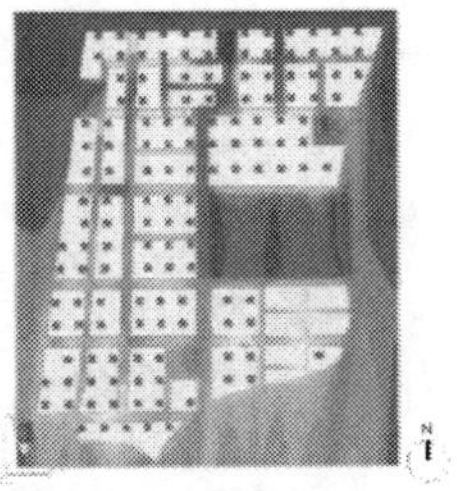

Figure 8.2.4 Hight_9M
Scenario III: 0.04 ha

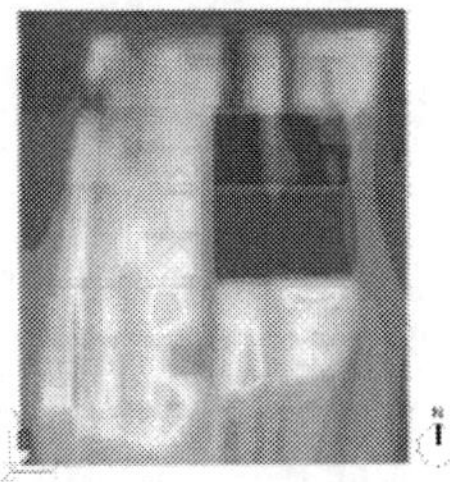

Figure 8.3.1 Hight_10M
Original plan: 6 ha.

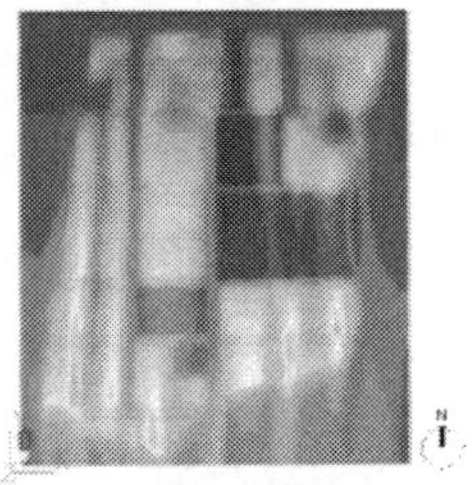

Figure 8.3.2 Hight_10M
Scenario I: 3 ha

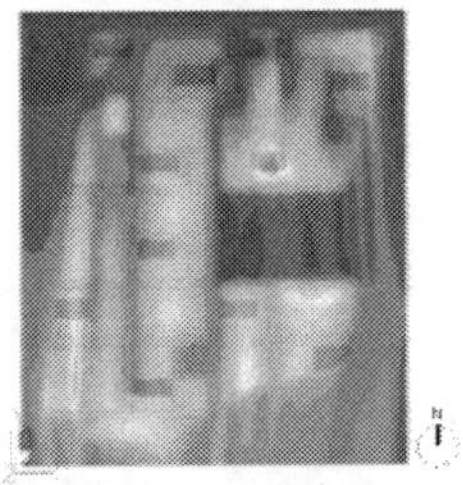

Figure 8.3.3 Hight_10M
Scenario II: 0.5 ha

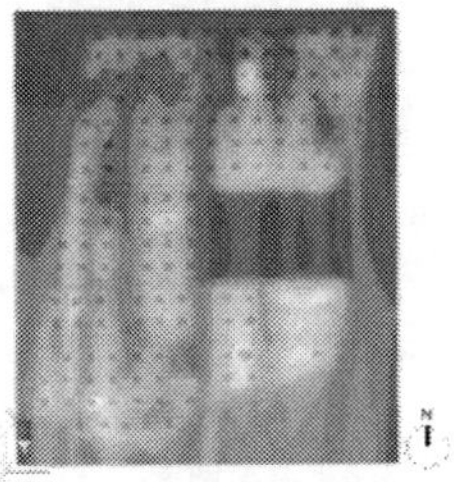

Figure 8.3.4 Hight_10M
Scenario III: 0.04 ha

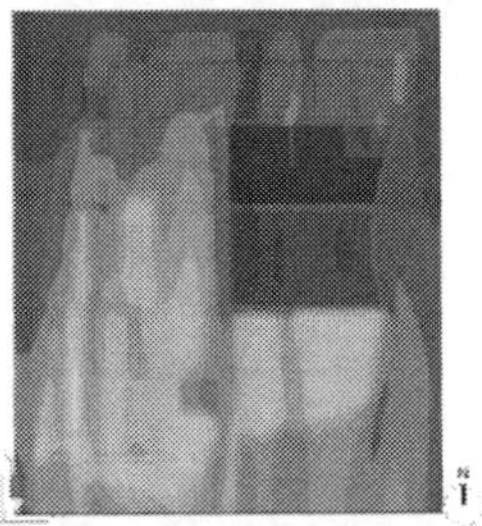

Figure 8.4.1 Hight_13M
Original plan: 6 ha.

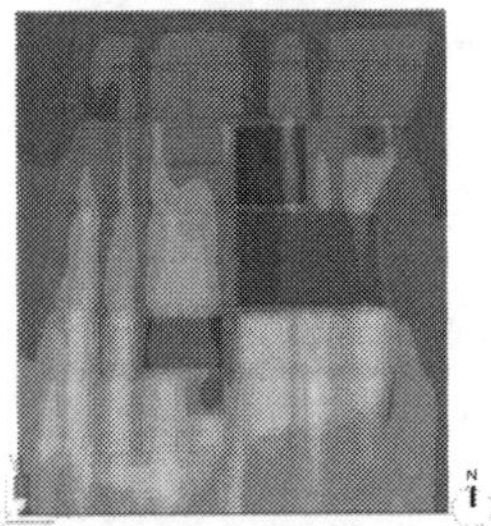

Figure 8.4.2 Hight_13M
Scenario I: 3 ha

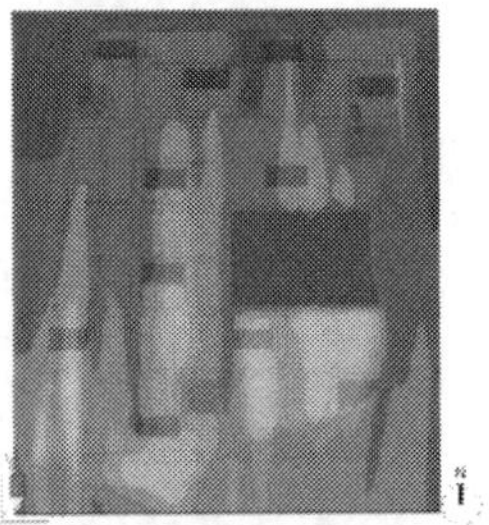

Figure 8.4.3 Hight_13M
Scenario II: 0.5 ha

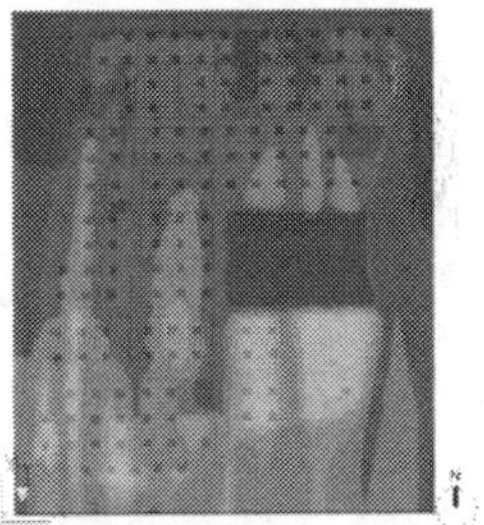

Figure 8.4.4 Hight_13M
Scenario III: 0.04 ha

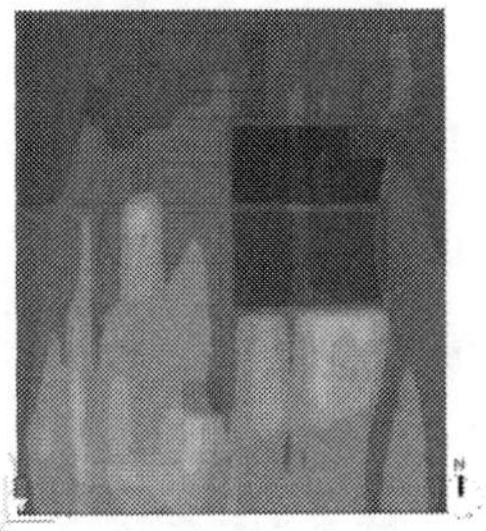

Figure 8.5.1 Hight_16M
Original plan: 6 ha.

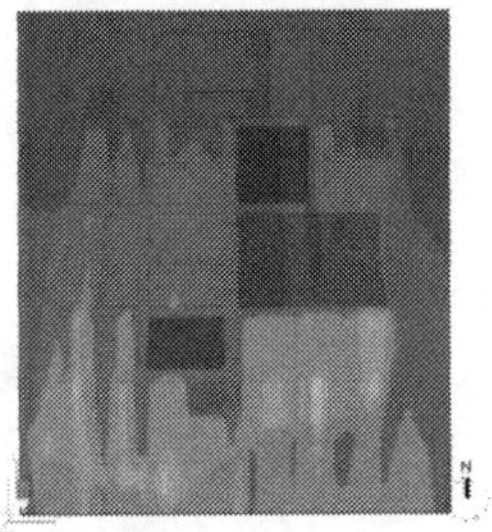

Figure 8.5.2 Hight_16M
Scenario I: 3 ha

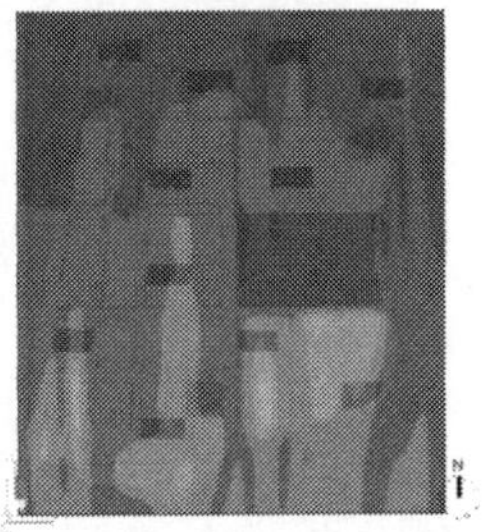

Figure 8.5.3 Hight_16M
Scenario II: 0.5 ha

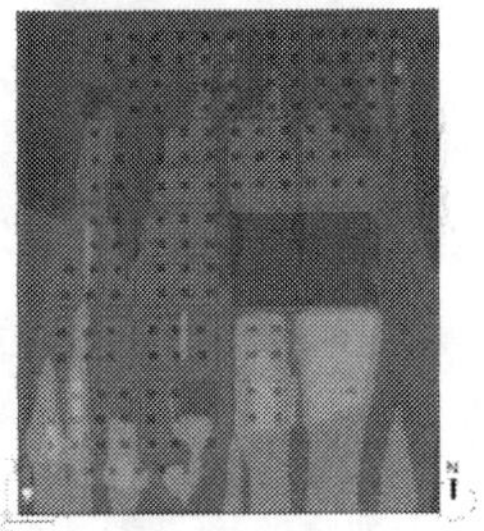

Figure 8.5.4 Hight_16M
Scenario III: 0.04 ha

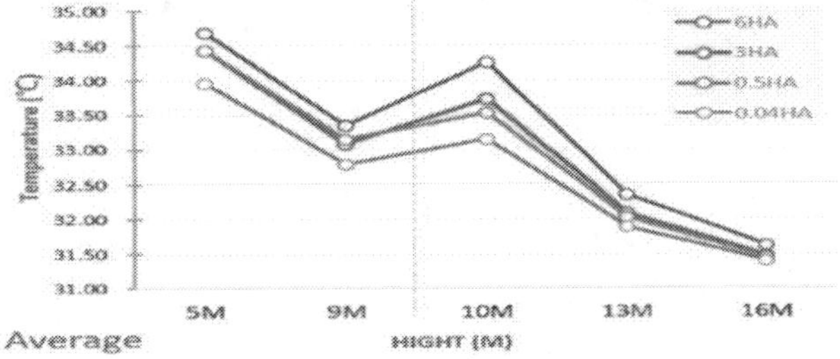

Figure 9. The Average temperature of 5 heights of 4 scenarios

References

1. C. T. Lin, 2016, Urban Forestry & Urban Greening Volume 20, Preliminary study of the influence of the spatial arrangement of urban parks on local temperature reduction, pp.348–357
2. M. H. Lee, 2004, A Study on the Relationship between Urban Parks and Local-climate - A Case Study in Taipei.
3. Wenqi Lin, Ting Yu, Xiangqi Chang, Weijia Wu, Yue Zhang, 2015, Calculating cooling extents of green parks using remote sensing: Method and test, Landscape and Urban Planning 134 (2015) 66–75
4. Construction and Planning Agency (CPAMI), 2011, Planning and Design Handbook of Park Greenbelt system.
5. Masatoshi Yoshino, 1976, 《Microclimate》, Daming Church Co., Ltd.
6. Boundary Layer Climate (Second Edition). By T. R. Oke. May Hugh. year 1987.

Eco-Design and Energy Saving Concepts for the Agricultural - Commercial Fringe Communities in Western Thailand

Pastraporn Thipayasothorn [1], *Rachadakorn* Phonpakdee[2], *Saravanee* Phuengpunum[3] and *Aumamporn* Pechsinjon[4]

[1]Assistant Professor with Faculty of Industrial Education and Technology, KMITL, Bangkok Thailand
[2]Associate Professor with Faculty of Industrial Education and Technology, KMITL, Bangkok Thailand
[3]Assistant Professor with Faculty of Administration and management, KMITL, Bangkok Thailand
[4]Administrator with Faculty of Industrial Education and Technology, KMITL, Bangkok Thailand

Abstract. These The purpose of this study was to develop planning for the communities along the agricultural -commercial fringe in the Special Economic Zones of western Thailand corresponding to the area's physical environment, activities and commuting patterns .In order to fulfill the purpose, this study has analyzed the significant activities and commuting patterns of the communities along the particular fringe area in the Special Economic Zones of 5 western provinces as well as the opinions towards community planning given by 400 residents living in the certain areas. The findings indicated that the establishment of an economic development plan for the development of commuting routes of the Special Economic Zone in the western part of Thailand is very appropriate. The plan will ultimately enhance economic development by optimizing all commuting routes, activities, and resources that originally exists in the local environment and economically impact all community throughout every region of Thailand.

Keywords. Eco-Design planning, Energy saving concept, Agricultural - commercial fringe area

1 Introduction

Soon after Thailand has entered into the ASEAN Economic Community (AEC) in 2015, international trade among ASEAN countries has grown rapidly [1]. According to the report of the Thai Board of Investment (BOI), the industrial market zone has been divided into six sectors that are North, Northeast, Central, Eastern, Southern, and Western.

In order to prepare Thailand's industry to be ready for the ASEAN Economic Community in 2015, the Industrial Estate Authority of Thailand (IEAT) had launched the project to develop Thailand's border areas in many provinces consisting of Mukdahan, Songkhla, Sa Kaeo, Trat, Kanchanaburi, and Tak. The project was expected to support the expansion of the industry, rise of hiring rates, and numbers of emerging factories, especially small and medium enterprises (SMEs) that became an important business group for each ASEAN countries.

The expansion of the industry in many areas has direct impacts on the area as land-use was transformed from agriculture to industry. For instance, the communities located around the industrial estate have to face with environmental issues occurred due to the lack of social responsibility and environmental awareness of many industrial estates operating across the country such as Map Ta Phut Industrial Estate in Rayong. In the past, the majority of the population residing in the surrounding villages earned their livings by farming and fishing. Decentralization according to the 5th National Economic and Social Development Plan led to the implementation of industrial estates and deep-sea port that has an impact on the communities around them. For example, various kinds of toxic substance were released and polluted the environment of the communities, schools, and residences, close to each industrial estate.

Dawei Deep Seaport and Industrial Estate Development Project in the Special Economic Zone has caused changes in the socio-economic system, resulting in a large number of employment, country's infrastructure, transportation, and other facilities have been increasing. However, all the advancement happened be the industrial estates might follow by both direct and indirect problems to the community if they were carried without appropriate management planning [2]. Therefore, the approach to sustainable development and reduce the impact on communities around the industrial estates required research toward infrastructure planning for any emerging industrial estate to be environmental friendly and effectively decrease the impact on the community and the environment.

Economic growth has been a major reason for the spread of urban sprawl that invaded into nature and agriculture and caused a permanent loss in sources of food production and water retention. Every minute the city sprawls, the greater loss it becomes for the agricultural land. Every square meter of farmland lost is equal to the increase in food prices. Besides, the more agricultural area was pushed further from the city, the fewer opportunities for urban residents to reach the quality and healthy food. Although we are happy with the economic growth, we have to stop the urban sprawl from invading the agricultural land for food security and quality of life [3].

Apart from that matter, there is also an urgent need for a lot of improvement. The remaining agricultural areas must be restored to be the food supply for the people. The education on value-added agriculture and increasing production productivity to develop the export potential in the groups of product with high export demands are very important as it helps generate revenue for the community. Moreover, the education on value added agriculture will further the knowledge and commercially benefits the local community and the public. In addition, the development of product quality regarding the Food Safety and Food Security Standard is a necessary part of the urgently required improvement as well.

Since the present Western Special Economic Zone is still lack of clear composition to maximize the benefits for the people, this research aims to study the development of plan for the commuting routes in the 5 western provinces, namely; Kanchanaburi, Tak, Prachuap Khiri Khan, Phetchaburi, and Ratchaburi, together with the survey on demands of the local people. Therefore, the research concentrated on approach to support the expansion of the industry and link all regions of the country together for the sustainability of the communities along the agricultural - commercial fringe in the Special Economic Zones of western provinces in Thailand.

2. Objective

1. To study the physical characteristics of the community in the Special Economic Zone, Western Region
2. To study the composition in the community plan of green communications and energy saving of ones along the agricultural - commercial fringe in the Western Special Economic Zone
3. To propose a community planning scheme for the communities along the agricultural - commercial fringe in the Western Special Economic Zone

3. Scope of Study

The scope of the study of planning for the communities along the agricultural - commercial fringe for the sustainability of the Special Economic Zones in western Thailand is to study the physical characteristics of the communities, natural resources, arts, culture and traditions in the Special Economic Zone in the particular western provinces. Additionally, to propose a community planning scheme for the communities along the agricultural - commercial fringe in the western Special Economic Zone, studying the composition in the community plan of green communications and energy saving is also needed. The scope of study can be explained as follows.

3.1 Population and Sampling Scope

1. Population: 2,814,370 residents with 20 years of age and over in the 5 western provinces consisting of Kanchanaburi, Tak, Prachuap Khiri Khan, Phetchaburi and Ratchaburi
2. Sample: 500 participants were selected from all the population living and doing activities the 5 western provinces; 400 residents and 100 government officials taking charge of agricultural-commercial promotion and community leaders in the 5 western provinces

3.2 Content Scope

1. Physical characteristics of the community in the Special Economic Zone, western region
2. Composition in communities along the agricultural - commercial fringe in the western Special Economic Zone
3. Planning for the communities along the agricultural - commercial fringe in the western Special Economic Zones

Kerdmemoon has influenced the conceptual framework of physical planning for the communities [4] along the agricultural-commercial fringe, facility needs and local issues. In terms of location, he pointed the condition of current usage, public utilities and facilities including the uniqueness and lifestyle of the community are the physical characteristics of each community. It is an impression of the people who have seen the total picture of the communities and how they recognize them after visiting the area. These simply refer to how the image of the communities can be imprinted into people's mind.

3.3 Area Scope

The studied areas include the western region of Thailand which most part is basically an alluvial plain area so most people in the central western can earn their living from both freshwater and saltwater fishing. Trading is also popular among Thai westerners because they have very convenient land-water transportation great for trade. The western region consists of 5 provinces that are Kanchanaburi, Tak, Prachuap Khiri Khan, Phetchaburi, and Ratchaburi.

4. Literature Review

Regarding the theories related to local community, the term 'local community' refers to the place where people connecting to a geographic environment and social relation system. The social that based on generosity, engagement and interactions between the locals is an important method to make a good life together.

In other words, the local community also refers to a combination of people who spatially and socially have a long-standing relationship. Thus, this is the basic unit of self-reliance and self-management and it involves participation in social drive and a psycho-cultural unit that identifies the community in relation to ideology, rights, and authority. In addition, the meaning of the local community according to the sociological concept refers to the social and physical unit such as neighborhood areas, village, and city. All of these share meanings as follows.

1. Local community as a territory unit having a concrete nature of the membership and the stable settled by geographical area

2. Local community as a unit / social system being a network of human interaction that correlates status, role, group of people, and community institution together in chains

3. Communities as a unit of cultural psychology focusing on the bond between community's members either psychologically or culturally

The meaning of the local community based on the concept of anthropology concentrates on the community as a cause of friendship, generosity, security, and commitment. Intellectuals who support this concept proposed that there was a need for small communities with a tight structure as it will help restoring social well-being and harmony. This concept reflects the dynamics of modern society, the expansion of the city that caused controversy and ignorance among the ideal community. Consequently, the sense of community will appear when people express their individualities, values, and relationships of those who live together. According to the concept of Kanchanapa, the main characteristics of community are composed of four factors which include;

1. Social and cultural values that are the value of generosity, interdependence, and honesty

2. The social capital that is a mindset toward knowledge management system in the community such as public resource usage and the arrangement of coexistent relationships, etc.

3. Righteousness which refers to rights everyone from every social class deserves including justice administration operated under the horizontal and vertical social relationships

4. Dynamic Learning focusing on learning and self-adjustment to create a strong and solid community that contains sense of unity and the ideology relating to rights in administration, adaptation and learning processes of some particular area to meet the needs of local people according to the horizontal and vertical social relationships.

As mentioned above, these community characteristics must share the same of components which are people, places or areas, social interest, social interaction, social relationship, social organization as well as social benefits. Boonplook [5] described the nature of the local community as an organization or social entity that is composed of 5 related and interrelated components as follows; 1. Having size based on human capabilities 2.Having a sense of identity 3.having a sense of possession and 4.having a bond with a rural culturally oriented society 5.Being close to a rural culturally oriented society.

The western part of the region is mostly mountainous, from both old and new eras. This mountainous area is the resources of mineral, forests, and watersheds but is not suitable for agriculture as it has less space. In addition, the mountains below this region are also located in the north-south direction forming a barrier of wind and humidity from the Andaman Sea bringing less rainfall to the area of downwind. However, when the government has built a dam for irrigation and established the irrigation canal system for the eastern plains of the region[6]. The dam makes this area an important agricultural plain even though it is a rain shadow area. Since the western region is not far from the capital and it contains numbers of scenic waterfalls, rapids, coasts, shallow sea, islands and sandy beaches attracting many tourists to this region annually, the locals can create income from the above-mentioned advantages that encourage them to do agricultural profession.

5. Methodology

The study of planning for the communities along the agricultural - commercial fringe for the sustainability of the Special Economic Zones in western Thailand is non-experimental research carried out by the following methods;

1. The physical characteristics of the community in the Special Economic Zone, western Region

2. The composition in the community plan of green communications and energy saving of ones along the agricultural - commercial fringe in the western Special Economic Zone

3. Propose of the community planning scheme for the communities along the agricultural - commercial fringe in the western Special Economic Zone

– Population participating in this research consisted of 2,814,370 residents with 20 years of age and over in the 5 western provinces consisting of Kanchanaburi, Tak, Prachuap Khiri Khan, Phetchaburi and Ratchaburi

– Sample of this research was 500 participants selected from all the population living and doing activities the 5 western provinces; 400 residents and 100 government officials taking charge of agricultural-commercial promotion and community leaders in the 5 western provinces

4. The data collection instruments

- Quantitative data collection: the author had conducted survey and observation in a structured way by designing and planning the following procedures;

a)Collect all printed documents, maps, photos and data from related websites.

b)Draft case study observation covering the physical characteristics of the western region.

c)Examine survey instruments and observation given by 2experts.

d)Survey and observe the western region of Thailand.

e)Summarize results of the survey and observations.

The data from the survey and observations above became the physical data that the author analyzed with the data from the first part of the questionnaire according to the research process.

- Questionnaires: the selected participants from 5 western provinces were asked to answer the questionnaires. The author had studied and designed methods for data collection as the following procedures;

a) Study data from printed documents, publications, websites and research related to the physical characteristics of the western region.

b)Study the design of community plan of green communications and energy saving for the communities along the agricultural - commercial fringe in the western Special Economic Zone

c) Draft the questionnaire according to the purpose of the study to ask opinions of the sample in the studied area. The questionnaire consists of 4 parts; 1. General information of the respondents 2. Physical characteristics of the western region 3. Components of the community plan of green communications and energy saving of the stated area 4. Planning scheme for the communities along the agricultural - commercial fringe in the western Special Economic Zone.

d)Consult with 3 qualified experts on the questionnaire for appropriate adjustments, and then carry a pilot survey in the non-sample area outside the study area.

e) Adjust the questionnaire to be suitable for data collection.

f) Collect data with completed questionnaire in the study area.

g)Summarize the results of the questionnaire.

The information obtained from the survey questionnaire was personal opinions of the sample toward the planning of the communities along the agricultural - commercial fringe in the western Special Economic Zone. The certain data was analyzed with the survey and observation data following the research process.

- Data collected location: Thailand's western region

5. The Qualitative Data Analysis used related computer programs such as Adobe Photoshop, Microsoft Excel, and 3D Auto Cad in order to process the data collected from the survey and observations on facilities of community in the 5 western provinces. The Quantitative Data Analysis also used the mentioned computer programs to process the data collected from the questionnaire.

6. Summarize the community planning scheme for the communities along the agricultural - commercial fringe in the Western Special Economic Zone with the basic information derived from the certain study.

6. The Result

Planning for sustainable community lying between agricultural and commercial research in the Special Economic Zone of western Thailand was run by studying the sample population groups in the sample areas. We used the questionnaire and observation forms to be our instruments, then analyze the data according to the research's processes by dividing them into the following topics;

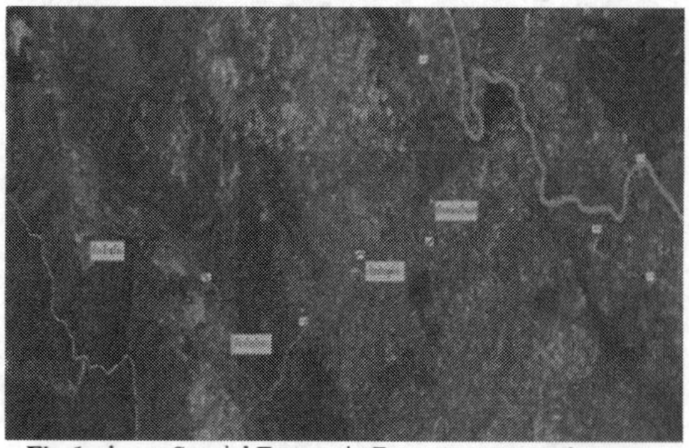

Fig. 1. shows Special Economic Zone, western part's route
Source: www.google earth.com

The study of the communities' physical characteristics in the Special Economic Zone, western part according to research objectives 1: As the data analysis from the surveys and observations, the west is not far from the capital. There are numerous waterfalls, islands, coast, reservoirs, shallow waters and sandy beaches. This is today, there are many beautiful sceneries, so it attracts many tourists which earn very good income. From the above mentioned environment, it is a factor to promote the local people to get various occupations according to the physical characteristics as shown in Figure 1.

From Figure 1, it was found that the current physical condition of the road leading to the village is ascent. The main road is a concrete road, and the road to Burma is constructing. The ways outside the residential areas which are along by the forest and agricultural crops have no street light.

From the link between the agriculture and commerce in the community. The western part is a border to Myanmar, so they do trading at the border which is areas of Tak and Kanchanaburi. Major products are consumer goods, processed timber, jewelry and forest materials. In a part of tourism service, this region has many tourist attractions such as waterfalls, jungles, reservoirs, islands, river, and beach. It is a popular tourist destination, especially in Kanchanaburi, Phetchaburi and Prachuabkirikhan.

Communities' economic development planning guideline in the Special Economic Zone, western region according to research objectives 2: Analyze questionnaire data in physical needs of the local communities and visitors to propose a layout of the Special Economic Zone as shown in Table 1.

Table 1. the relationship between the length of living time of the local people and the facilities they need

Length of living time	The facilities need				
	Hotel	Gas station	Market	Police station	Total
1.Less than 5 years	0	4	8	4	16
2.5 - 10 years	0	8	8	4	20
3.11 – 15 years	12	4	16	8	40
4.16 - 20 years	12	12	24	20	68
5.21 - 25 years	12	44	48	10	114
6.26 - 30 years	4	8	16	20	48
7.More than 30 years	0	8	64	22	94
Total	40	88	144	88	400

From Table 1, it is found that those who has been living in the areas about 25-21years prefer to have a department store the most. Secondly, those who has been living in the areas about 30years prefer a security center such as police station the most.

To propose a planning for communities lying between agricultural and commercial guideline in the

 https://doi.org/10.1051/e3sconf/20185704003

Special Economic Zone, western part according to research objectives 3: Private section support. To develop the commercial system to be higher and expand the products and services distribution both domestic and international *markets. Also, promote Thai people to be more interested* in trading and service jobs. In addition, the state need to maintain the prices for domestic consumption which is a key factor in the cost living and economic indicator.

7. Discussion

The study of the communities' physical characteristics in the Special Economic Zone, western region according to research objectives .1

According to the physical characteristics, the areas mostly are agricultural areas which are located in a distance from each other. This is consistent with the communities' economic development planning guideline in the Special Economic Zone, western region [7]. According to the needed facilities observation, the trading routes and transportation in Ban Muang District, Amphoe Mueang, Kanchanaburi are convenient [8]. Communities' economic development planning guideline in the Special Economic Zone, western region according to research objectives :2According to the study, the current areas have no place to be a main center for economic development which is contrast to the importance and readiness of Thailand concept [9]. According to the basic infrastructure development guideline, we need to solve community problems, and improve trading and tourism. Including, develop the core infrastructure follow the Border City Development Action Plan as well. It also consistent with the concept of free trade in Thailand.

It is believed that when the drive increases, the trade will expand by at least %25in some industries such as automobiles, tourism and transportation.

To propose a planning for communities lying between agricultural and commercial guideline in the Special Economic Zone, western region. According to the research objectives 3, the trade in the western Special Economic Zone will expand more which resulted to the capital sources[10]. Thus, the capital sources will be emerging such as hotel, tourism, restaurants and car rental industry. As there will be more roaming route and the communities lying between agricultural and commercial will develop more, the boarder areas will grow up more. The reason is there is a central of transportation hub in western part.

8. Suggestion and Recommendation

1. The results of planning for sustainable communities lying between agricultural and commercial in the Special Economic Zone in western Thailand will be three issues; physical characteristics, community plan of green communications and energy saving elements' design and planning for communities lying between agricultural and commercial guideline. The recommendations can be classified as follows; 1. Physical characteristics of the community in the Special Economic Zone, western region.

2. Trading route and transportation in the zone. According to the study, there are constructing the trading route which is connected to Myanmar. Therefore, people in the areas want to have a career which is work in the trade route.

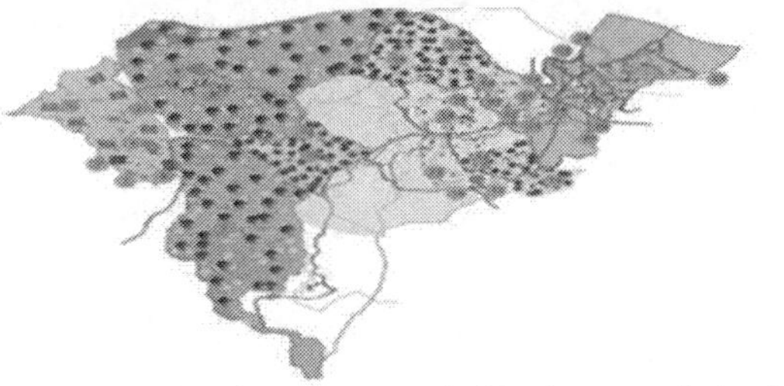

Fig. 2. shows planning for community lying between agricultural and commercial at Ban Kao district,Kanchanaburi

3. The physical characteristics of the communities on traffic patterns for trading and transportation. According to the study, they are constructing the trading route to connect with ASEAN countries.

Community plan of green communications and energy saving elements' design and planning for communities lying between agricultural and commercial guideline. Layout of space utilization by dividing areas into zones which are agricultural, residential, commercial, green, industrial areas as well as roaming routes.

Figure 3 shows the zone arrangement for using the layout design. The area is divided into 6zones to show the details of the designs in each zone.

4. Purpose the planning for communities lying between agricultural and commercial guideline. Layout of space utilization in different zones as well as roaming routes. The routes should be connected to the activities areas, and the commercial zone is given to locate in front of both sides of the Kwai Noi River which creates many activities as Ban Kao district, Kanchanaburi in Figure 2.

We study the plan for sustainable communities lying between agricultural and commercial in the Special Economic Zone of western Thailand to increase the value of agricultural products. Also, to develop export potential in agricultural products that generate very good income for the communities to commulate thier wisdom as well as quality development. To strengthen the compete ability and the communities' economy in order to earn income for the country and develop overall economy and socialment. It can build green communications and energy saving power in negotiating as well.

References

[1] Dokmai, Bancha, and Aekapol Jongwilaiwan. "Arrangement in World Economical Investment." WTO Watch, .2007

[2] Thipayasotorn, Pattraporn, and Jintana Nokyoo. "Coastal Area Development by Roaming Route Research." Bangpoo District Case Study, .2014

https://doi.org/10.1051/e3sconf/20185704003

[3] Department of Public Works and Town & Country Planning. "Bangkok's City Plan Guideline." Department of Public Works and Town & Country Planning, .2009
[4] Department of Public Works and Town & Country Planning. "City Plan of Thailand in ".2057 Department of Public Works and Town & Country Planning, .2008
[5] Boonplook, Orawan. "Mixing Agriculture according to New Theories." .2003
[6] Nicrowattanayingyong, S. Towards Good Governance in Town and Country Planning. Nonthaburi: Thailand Environment Institute, .2009
[7] Anussorn, Panna. "Food in Ayurveda Style: Healthy Food for New Generation." Good time, .2005
[8] Rodjanapradit, R. *Urban Logistics Planning*. Bangkok: Chulalongkorn University Press, .2007
[9] Rojanapradit, Rahaj. "City Planning for Logistic." .2007
[10] Department of Public Works and Town & Country Planning. *Thailand national spatial development plan 2057*. Bangkok: Ministry of the Interior, .2008

On Optimal Sizing of a Solar Thermal Building: A Novel Approach

Shahrzad Parsania

Architecture Department, Islamic Azad University - Kerman Branch, Iran

Abstract. Within the last decade, domestic energy management has gained a lot of attention. As the complexity of the solar thermal system in terms of the number of system components and energy sources increases, understanding how to manage the cooperation of all the components in order to improve the global efficiency measurements is of crucial importance. Here, the question is how to define an optimal size of the main components in a solar thermal system in order to minimize system cost. Unlike the existing approaches, we propose the use of a novel algorithm called Gravitational Search Algorithm (GSA) to analyze the accurate sizing of energy components, i.e. collector size, tank volume and Auxiliary Power Unit (APU). The objective is to maximize solar fraction, minimize the energy consumption and installation costs subject to constraints. Our proposed GSA model is evaluated and compared with one of the most well-known algorithms, Particle Swarm optimization (PSO) taking into account the fundamental system characteristics. Numerical results show that our proposed methodology significantly improves energy efficiency and reduces operational cost of the solar thermal system in contemporary built environment.

1 Introduction

With the advent of renewable energy, human beings have tried to overcome several severe limitations, including a limited supply of fossil fuels, high cost and carbon dioxide pollution. One of the most popular and fastest-growing sources of renewable energy is the solar energy which is available for local use in building heating, cooling, hot water supply and power production.

It also has the potential to meet most of the modern world energy demand by effectively integrating with local building components. Indeed, in order to phase out fossil fuels and reduce the related Co_2 emissions, solar energy is introduced as one of the most promising sources which can be used to provide hot water as well as environmental heating energy (a.k.a. solar combisystems [1]). These systems are, in fact, derived from the combination of common heating systems with solar collectors.

"Task 26" was a project initiated by the International Energy Agency (IEA) which conducts investigation on solar combisystems indicates that "if the direct use of solar energy is to make a significant contribution to the heat supply, it is necessary that solar-heating technologies must be developed and widely applied over and beyond the sole field of Domestic Hot Water (DHW) preparation"[2].

In particular, this research has focused on a solar combisystem installation that satisfy both DHW and Space Heating (SH) requirements at the same time. Solar combisystems are relatively complex and their performance depends on many parameters, such as, orientation and tilt angle of solar thermal collectors, weather conditions, DHW usage profiles and so on [3]. Therefore, optimization techniques are found to be an efficient approach to get the most out of solar thermal combisystems.

Existing approaches on the solar combisystems mostly demonstrated the impact of geographic location [4] and the use of different HW load profiles [5] on solar fraction and energy consumption of the system. Also, two popular optimization techniques being used lately are Genetic Algorithm (GA) and Particle Swarm Optimization (PSO). Authors in [6] used GA technique in order to minimize the payback time of a solar domestic hot water heating system. Authors in [7] also investigated how to optimize large solar domestic hot water systems with respect to solar heat cost. Furthermore, in [2], authors applied PSO to discuss the optimal sizing of the main components of a solar thermal system in order to provide heat for both DHW and SH needs.

However, unlike the existing approaches, in this paper, we propose a novel approach which, according to a comparative study in [8], outperforms other well-known evolutionary algorithms including Particle Swarm Optimization [9], and Genetic Algorithm (GA) [10].

Hence, in this paper, we consider a typical mid-sized house located in Kerman, Iran, and propose a methodology to find the optimal component sizes for a solar combisystem. The simulation architecture is conducted via TRNSYS and MATLAB, using

MATLAB as the main software that calls TRNSYS for the iterations.

To summarize, the primary contributions of this work are as follows:

- We formulate a mathematical model, propose an optimization framework with the aim of minimizing the system cost in terms of energy consumption, installation cost and maximum solar fraction, taking into account collector size, storage tank volume and Auxiliary Power Unit (APU).
- We provide the main differences between GSA and PSO, encode the framework into a GSA model and define its fitness function to minimize the system objective.
- Finally, an evaluation of the proposed model is conducted by extensive simulations and compared with one of the current state-of-the-art evolutionary algorithms, PSO, to demonstrate how accuracy and system convergence of our algorithm affect the efficiency of domestic energy management in solar combisystems.

The rest of the paper is organized as follows. In Section 2, we describe a typical combisystem model and its assumptions. In Section 3, we formulate the problem. In Section 4, we present the proposed GSA optimization framework aligned with its all essential parameters. Section 5 evaluates the simulation results, and, finally, Section 6 concludes the paper.

2 A solar thermal combisystem model

Designing a solar thermal combisystem implies, first, selecting a configuration, and then sizing all its main components. Fig. 1 shows the setup scheme of the solar thermal combisystem used in this research. This residential solar combisystem, proposed in [4], will be used as our initial design solution, where the multi-objective optimization framework is used for the selection of a configuration and its main components sizing. However, seven out of the 21 generic solar combisystem configurations considered by "Task 26" utilize more than one storage tank, most of the related works make an assumption of only one tank [3]. Thus, in this research, we follow Lund et. al. [4] and assume also hot water within storage tanks for space heating requirements too.

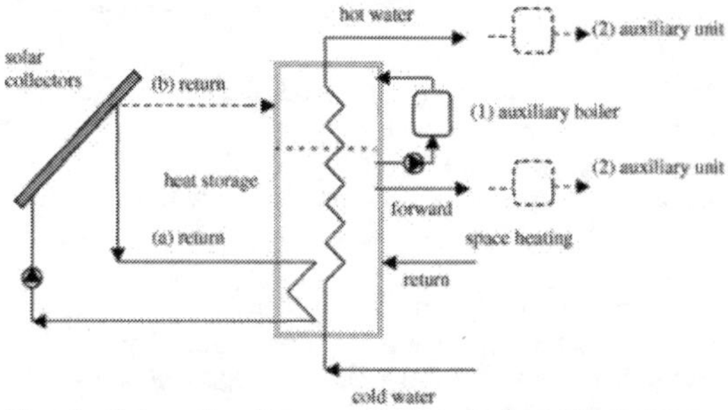

Fig. 1. Schematic diagram of a typical residential solar combisystem [4]

Both solar collector and APU are associated to this storage tank. The loads are specified by the DHW and SH demands. The load profile for DHW is generated as in [5], which considers a mean consumption of $200\ l/d$ and a load temperature of 45°C.

In fact, providing energy for space heating is a bit more difficult to determine.

Because it is correlated to the specific ambient temperature and building insulation. In our area of interest, the average temperature is collected from Worldwide Meteorological Organization (WMO: 408410) with Latitude and Longitude specified as $30.25N$ and $56.97E$ respectively. Also, the energy consumption of a typical residential building during a period of one year is collected and provided as in Fig. 2.

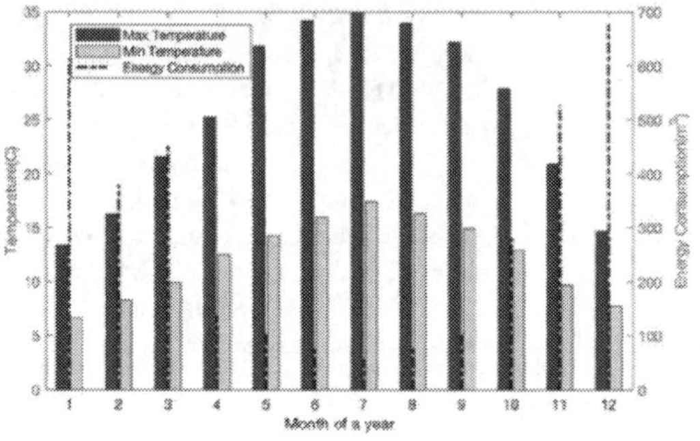

Fig. 2. Temperature and Energy consumption analysis for the area of interest.

Without loss of generality, for any given area of interest, since the precision of the simulations depends on the availability of realistic data of solar system parameters, e.g. irradiation and humidity, the collected monthly mean values of data from the closest weather stations from WMO, illustrated in Table 1, can be represented more accurately in hourly manner with the same statistical properties as the measured data [11].

Table 1. Kerman Radiation Data (WMO: 408410)

Month	R_b	R_d
Jan.	2.411	0.351
Feb.	2.246	0.387
Mar.	2.154	0.409
Apr.	1.978	0.456
May	1.860	0.489
Jun.	1.894	0.480
Jul.	1.999	0.457
Aug.	2.116	0.426
Sept.	2.178	0.405
Oct.	2.267	0.390
Nov.	2.364	0.368
Dec.	2.403	0.358

Lat:30.25N Long:56.97E Elev:1754 StdP: 81.95,
WMO: World Meteorological Organizatio,
R_b: Avg. beam radiation (MJ/m^2h),
R_d: Avg. diffuse radiation (MJ/m^2h).

Table 2. Relevant system pricing factors

Item	Value	Unit
Collector price	400	$\$/m^2$
Storage Tank price	5.30	$\$/l$
APU price	62	$\$ \ kW$

3 Problem formulation

Generally, every optimization problem formulation is composed of at least three components: (i) objective functions, (ii) a set of decision variables, and (iii) a set of constraints for the objective functions and decision variables. More particularly, in our case, we intend to determine an optimal solar combisystem design by finding the optimum feasible sets of values of decision variables which, in turn, minimizes the function vector $F(x)$, while taking all the constraints into account. The goal is to minimize a general function $f: R^n \to R$ with respect to system parameters which can be achieved as follows:

$$\underset{x \in S}{\text{minimize}} \quad \alpha.f_1(x) + \beta.f_2(x) + \gamma.f_3(x) \tag{1a}$$

$$\text{subject to} \quad l^j \leq x^j \leq u^j \quad \forall j \in \{1, ..., n\}, \tag{1b}$$

$$\alpha, \beta, \gamma \in \mathbb{R} \tag{1c}$$

where $F(x) = [f_1(x), f_2(x), f_3(x)]$ and $X = \{x \subset \square^n \sqsubset l^j \leq x^j \leq u^j, j \in \{1, ..., n\}\}$ represent the objective function vector and the decision vector respectively; α, β and γ are weighting factors applied to the objective functions in order to adjust the importance of each objective; S indicates the decision search space; n denotes the number of decision variables where in our model $n = 3$, and, l^j and u^j identify the lower and upper bound of the j-th decision variable, respectively.

For instance, assuming collector size as a decision variable, we can define the lower and upper bound for this variable between 1 and 40 (List of input variables is provided based on [2] in Table 3).

Determining an optimal solar combisystem design implies finding a set of values of the decision variables $x^* \in S$ which optimizes the multi-objective function vector $F(x)$, taking into account all the constraints. In order to capture all the related quantities in the cost function, our objective vector is defined to maximize the solar fraction $f_1(x)$, to minimize the total amount of consumed energy $f_2(x)$, and to minimize the installation cost $f_3(x)$. It is worth noting that to maximize an objective function by a minimization function, we can simply choose a negative value for the weighting factor α.

Table 3. Input variable set

Variable	Min	Max	Unit
Collector size	1	40	m^2
Tank volume	100	3000	l
APU size	5	50	kW

Our proposed methodology is evaluated based on the GSA meta-heuristic optimization formulation [12]. GSA was originally derived from Newton's theory of gravity and mass interactions to solve optimization problems. According to the comparative study [8], the authors did show that GSA performs better than other well-known evolutionary algorithms including Particle Swarm Optimization (PSO) [13], and Genetic Algorithm (GA) [10]. In the following, we discuss how the algorithm works and then, evaluate the proposed methodology.

4 The Proposed methodology

To evaluate the performance of the proposed approach, we first explain how GSA algorithm operates; then, we discuss how effectively our approach responds to the system variables in the solar combisystems.

4.1 Background on GSA Modelling

In this section, the structure and main principles of GSA [12] are briefly described. Recently, GSA has gained more popularity due to its robustness and effectiveness (e.g. [8],[14]).

It was originally derived from Newton's theory of gravity and mass interactions to solve optimization problems. GSA is made up of particles (objects) that interact with each other by the gravity force. The performance metric of particles are measured by their masses. Particles try to move forward towards the heavier particle's mass under the gravity force law. This force pushes the global movement of all particles. Each particle indicates a solution to the problem. Thus, the heavier the particle mass is, the better the solution will be. For the algorithm, in order to work properly, two parameters, gravitational and inertia masses, should be set with appropriate values. While executing GSA algorithm, during iterations, particles get attracted to the heaviest particle mass. At the end, the heaviest particle in the search space represents an optimal solution to the problem. More precisely, suppose a system with n masses in which the position of i th mass is presented as follows:

$$X_i = (x_i^1, \ldots, x_i^d, \ldots, x_i^n)$$

where x_i^d indicates the position of i th mass in the d th dimension. The force of each mass j acting on mass i at time t can be calculated as:

$$F_{ij}^d(t) = G(t)\frac{M_{pi}(t) \times M_{aj}(t)}{R_{ij}(t) + \varepsilon}(x_j^d(t) - x_i^d(t))$$

where $G(t)$ represents gravitational constant at time t, $M_{pi}(t)$ and $M_{aj}(t)$ represent passive and active gravitational masses of agents i and j respectively. ε indicates a small constant, and $R_{ij}(t)$ indicates the Euclidean distance between masses i and j.

In order to add stochastic characteristic, GSA multiplies the sum of all forces exerted from other masses with a random number in the range of [0,1] as follows:

$$F_i^d(t) = \sum_{j=1, j\neq i}^{n} rand(j) \times F_{ij}^d(t)$$

By using the total force, acceleration of each mass i at time t can be measured by the law of motion:

$$a_i^d(t) = \frac{F_i^d(t)}{M_i(t)}$$

where $M_i(t)$ indicates the inertia mass of i. Moreover, based on the law of motion, mass dynamics consist of velocity v and position x. The velocity of mass i at time $t+1$ is measured as the fraction of velocity at time t, plus its acceleration, which can be formulated as:

$$v_i^d(t+1) = rand(i) \times v_i^d(t) + a_i^d(t)$$

Also, its position at time $t+1$ is calculated as:

$$x_i^d(t+1) = x_i^d(t) + v_i^d(t+1)$$

The gravitational constant $G(t)$ is a function initialized at the beginning of the algorithm and decreased over the time lapse to control search accuracy. It is defined as:

$$G(t) = G_0 \times exp(-\gamma \times \frac{iter_{current}}{iter_{max}})$$

where G_0 is the intial value and $-\gamma$ is the constant control parameter. $iter_{current}$ and $iter_{max}$ denote the current and maximum number of iterations, respectively.

Notice that mass value has direct relation with the mass attractiveness. Heavier masses are more attractive than the lighter ones; thus, they move slower and absorb lighter masses to themselves. Considering the fitness value of mass i at time t denoted by $fit_i(t)$, the following equations are used to update the gravitational and inertial masses:

$$m_i(t) = \frac{fit_i(t) - worst(t)}{best(t) - worst(t)}$$

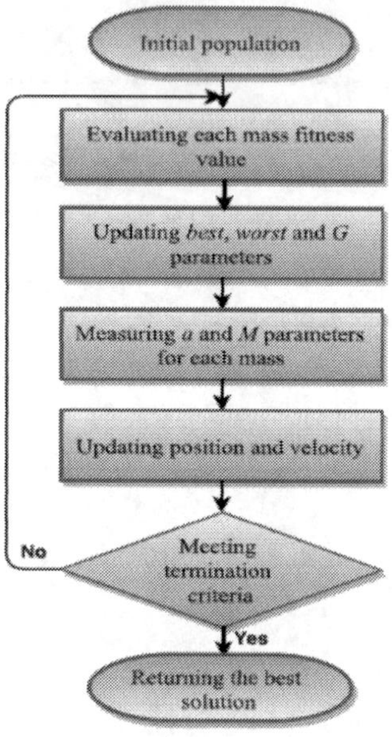

Fig. 1. GSA Algorithm Procedure

$$M_i(t) = \frac{m_i(t)}{\sum_{j=1}^{n} m_j(t)}$$

where for a minimization problem, $best(t)$ and $worst(t)$ are defined as:

$$best(t) = \min_{j\in\{1,\ldots,N\}} fit_j(t)$$

$$worst(t) = \max_{j\in\{1,\ldots,N\}} fit_j(t)$$

For the sake of simplicity, we illustrate GSA procedures in Fig. 3.

4.2 Comparison of GSA with PSO

Particle Swarm Optimization (PSO) is one of the well-known meta-heuristic algorithms in optimization problems. We will show that our proposed methodology with GSA performs even better than PSO. In this section, in order to fully understand how GSA works, the main differences between GSA and PSO are identified as follows:

- However, in both algorithms, GSA and PSO, the optimization is acheived by the movement of masses in the search space, the movement startegy is different.
- In PSO, particle's direction is identified using only two position values, *pbest* and *gbest* ; while in GSA, the direction of a mass is identified by the total force achieved by all other masses.
- PSO does not take the quality of solutions and the fitness values into account during the update procedure, while the force in GSA is proportional to the fitness value.
- In PSO, updating is performed without taking into account the distance between solutions, while the force in GSA is reversely proportional to the distance between solutions.
- In order to update velocity, PSO consumes memory to store the values of *pbest* and *gbest* . However, GSA is a memory-less algorithm and takes into account only the current position of masses.

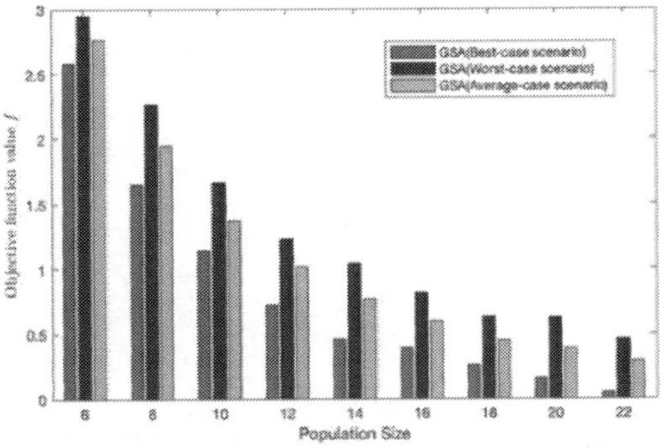

Fig. 2. Objective function vs. Population size

5 Experimental results

In this section, the effectiveness of the proposed approach is evaluated via simulation. The building solar thermal behavior is initially modeled by TRNSYS v.17 software.

Also, in order to technically evaluate the efficiency of GSA, we will compare our methodology with the PSO approach presented in [2]. Generic parameters of simulation are set as follows [12]. Maximum iteration is set to 1000. Initial value of gravitational constant G_0 is set to 100. In PSO, correction factors c_1 and c_2 are set to 2, and Inertia value (ω) is assigned with the value of 1. Plus, all the relevant installation factors for component pricing are selected from Table 2. For both algorithms, results are averaged over 50 runs and the average, best and worst solutions are also provided.

An important factor in implementation is population size. As the population size represents number of masses, larger population size results in a better solution. However, increasing population size impose more complexity to the system and, as a consequence, requires more time to complete the execution. To find an appropriate value for population size, Fig. 4 is provided, in which best, worst and average-case GSA solutions for population size of range 8 to 22 is measured. As illustrated, with the increase in population size, we notice better results in terms of system performance (i.e. the lower the better; as we are aiming to minimize the objective function). However, as population size increases, it adds up more complexity to the system. Hence, without loss of generality, we set the population size for all cases as 10. Also, to validate our approach, the average, best and worst solutions for the minimization function are compared in case of both GSA and PSO algorithms. As demonstrated in Table 4, our proposed GSA method achieves better results over 50 runs than PSO.

Table 4. GSA vs. PSO optimization of objective function values over 50 runs

Obj. Function	Best	Worst	Average
F_{GSA}^{opt}	1.138	1.667	1.370
F_{PSO}^{opt}	1.376	2.117	1.685

Investigations reveal that the average case for optimal cost function is reduced by over 20% for GSA at a faster convergence rate with higher degree of accuracy as compared with PSO. This preliminary results highlight the potential of this methodology, thereby improving the domestic energy management system in solar thermal combisystems.

6 Conclusion

In this paper, we targeted optimal sizing problem for solar combisystem and defined an effective optimization model to answer the fundamental question of finding the optimal size of the main components of a solar thermal system to go one step further towards efficient energy management. The analysis was performed based on a solar thermal combisystem for a typical mid-sized residential building in Kerman, Iran. We proposed the use of GSA algorithm to analyze the accurate sizing of energy components, i.e. collector size, tank volume and Auxiliary Power Unit (APU). The objective was to maximize solar fraction, minimize the energy consumption and installation cost subject to constraints. Our proposed GSA model was evaluated by comparing with one of the most well-known algorithms, Particel Swarm optimization (PSO) taking into account the fundamental system characteristics.

Numerical results demonstrated that our algorithm can significantly improve energy efficiency and reduce operational cost of the solar thermal system in contemporary built environment.

References

1. J. N. C. Hin and R. Zmeureanu, "Optimization of a residential solar combisystem for minimum life cycle cost, energy use and exergy destroyed," *Sol. Energy*, **100**, 102–113 (2014)
2. R. Bornatico, M. Pfeiffer, A. Witzig, and L. Guzzella, "Optimal sizing of a solar thermal building installation using particle swarm optimization," *Energy*, **41**, 31–37(2012)
3. A. Rey and R. Zmeureanu, "Multi-objective optimization of a residential solar thermal combisystem," *Sol. Energy*, **139**, 622–632 (2016)
4. P. D. Lund, "Sizing and applicability considerations of solar combisystems," *Sol. Energy*, **78**, 59–71 (2005)
5. U. Jordan and K. Vajen, "Influence Of The DHW Load Profile On The Fractional Energy Savings:: A Case Study Of A Solar Combi-System With TRNSYS Simulations," *Sol. Energy*, **69**, 197–208 (2001)
6. M. Loomans and H. Visser, "Application of the genetic algorithm for optimisation of large solar hot water systems," *Sol. Energy*, **72**, 427–439 (2002)
7. M. Krause, K. Vajen, F. Wiese, and H. Ackermann, "Investigations on optimizing large solar thermal systems," *Sol. Energy*, **73**, 217–225 (2002)
8. A. Zarrabi, E. K. Karuppiah, Y. K. Kok, N. C. Hai, and S. See, "Gravitational Search Algorithm Using CUDA," in *2014 15th International Conference on Parallel and Distributed Computing, Applications and Technologies*, 193–198 (2014)
9. M. A. Rodriguez and R. Buyya, "Deadline Based Resource Provisioningand Scheduling Algorithm for Scientific Workflows on Clouds," *IEEE Trans. Cloud Comput.*, **2**, 222–235(2014)
10. K. F. Man, K. S. Tang, and S. Kwong, "Genetic algorithms: concepts and applications [in engineering design]," *IEEE Trans. Ind. Electron.*, **43**, 519–534, (1996)
11. R. J. Aguiar, M. Collares-Pereira, and J. P. Conde, "Simple procedure for generating sequences of daily radiation values using a library of Markov transition matrices," *Sol. Energy*, **40**, 269–279(1988)
12. E. Rashedi, H. Nezamabadi-pour, and S. Saryazdi, "GSA: A Gravitational Search Algorithm," *Inf. Sci. (Ny).*, **179**, 2232–2248(2009)
13. J. Kennedy and R. Eberhart, "Particle swarm optimization," in *Neural Networks, 1995. Proceedings., IEEE International Conference on*, **4**, 1942–1948 (1995)
14. A. Kant, A. Sharma, S. Agarwal, and S. Chandra, "Swarm, Evolutionary, and Memetic Computing: First International Conference on Swarm, Evolutionary, and Memetic Computing, SEMCCO 2010, Chennai, India, December 16-18, 2010. Proceedings," B. K. Panigrahi, S. Das, P. N. Suganthan, and S. S. Dash, Eds. Berlin, Heidelberg: Springer Berlin Heidelberg, 286–295 (2010)

The Application of Economic Value Added on Green Facilities of Urban Agriculture

Kuo Wei Hsu[1] and *Shang Zhen* Liao[2]

[1]Professor, Urban Planning, Land Development, Chaoyang University of Technology,No.168, Jifeng E. Rd., Wufeng Dist., Taichung City 413, Taiwan (R.O.C.)

[2]Urban Planning, Land Development, Chaoyang University of Technology, No.168, Jifeng E. Rd., Wufeng Dist., Taichung City 413, Taiwan (R.O.C.)

Abstract. Green facilities of urban agriculture offer a fertile ground for green innovation, such as rooftop gardens, green curtains, and indoor plant growing, which create a new forms of urban farming economy. The previous studies for cost-benefit analysis of urban agriculture indicated that green facilities could contribute both quantitative and qualitative benefits. The value of green facility strategies of urban agriculture may often be underestimated without transforming qualitative benefits to quantitative benefits. This study examined the effects of green facilities of urban agriculture development in Taiwan. The market reaction was explored by adopting economic value added (EVA) technique for the measurement of both quantitative and qualitative value contributed by green facilities. The results from this study support that there exists interaction between qualitative factors such as indirect economic benefits and social psychological benefits and quantitative factors, agricultural production and energy saving impacted by green facilities. The study concluded that consumers of urban agricultural crops are willing to pay more for eco-friendly products. The establishment of relationship between urban agricultural production and consumption can contribute to sustainability while generating economic opportunities.

Keywords. Green Facilities, Urban Agriculture, Economic Value Added (EVA)

1 Introduction

Urban agriculture integrated with green facilities has been widely accepted in major metropolitan cities of the world. Urban agriculture offers not only agricultural products but also a platform for green innovation by linking its technology to the principles of sustainable urban environment. Green facilities of urban agriculture offer a fertile ground for green innovation, such as rooftop gardens, green curtains, and indoor plant growing, which create a new forms of urban farming economy. In recent years, Taiwan's public sector and non-profit organization have been pushing forward the urban agriculture policy and act. However, the previous studies for cost-benefit analysis of urban agriculture that had ignored transforming qualitative benefits to willingness to pay leaded to underestimate the influences of green facilities. Thus except rewarding green facilities of urban agriculture, to expose the real benefits can encourage private sector to participate the urban agriculture act. The objectives of this study were:

- To explore the interaction between qualitative and quantitative value-creation factors of green facilities of urban agriculture.
- To apply economic value added (EVA) technique on exploring the determining costs and benefits of urban agriculture associated with green facilities.

2 Literature Review

The urban expansion and the change in people's working style or habits led many young people choose to work in urban area in the recent years. However, the reduction of the population who cultivate and the traditional agricultural land' s usage amount led the culture of traditional agriculture disappear gradually. Agriculture has emerged in cities at that time. Urban agriculture develops different forms of urban agriculture with different urban issues and social backgrounds. It complies with the four conditions, purpose, space area, planting scale, and production technology to constitute the definition of urban agriculture. [1] The obvious performance of agricultural production and management is highly intensive mode of operation, and it achieve the production, processing and marketing of integrated management, and thus achieve a high degree of agricultural development and urban services for the special features. Most of the crops that are cultivated are crops of relatively high economic value with short production cycles. Most of the crops are sold to local residents or sold to nearby restaurants through self-production and sales, using local markets for sale. Urban agriculture can offer benefits as follows: 1. source of fresh and safe products, 2. opportunity for urban residents' engagement in agricultural activities, 3. open space for disaster management, 4. resource for recreation and well-being, and 5. education and awareness-raising for improving urban residents' understanding of agriculture and food issues. Urban agriculture can contribute to sustainability and well-being in cities by increasing the area of permeable surface for storm water management, or reducing the heat-island effect by cooling the air. Through green meeting places in neighbourhoods, group-based nature activities (e.g. walking or willow pollarding) and gardening (shared gardens for the elderly and allotment gardens)[2] ,urban agriculture also can contribute to biodiversity and ecosystem services by providing habitats and managing species. Furthermore, it can reduce food miles and even provide bio-energy resources.

3 Taiwan's Green Facilities of Urban Agriculture Experience

The most common green facilities of urban agriculture are planting plants on the roof and public open space with fruits, vegetables, and herbs. The type of facilities in urban agriculture in the neighborhood can build a more complete industrial structure. [3] Green facilities of urban agriculture achieved the green, leisure and production of multiple functions. Building plants on the roof also provides many ecological and economic benefits. [4] Green roofs have a layer of living plants on top of the structure and the waterproofing elements. Green roofs can be installed on a wide range of buildings, from industrial facilities to private residences. Figure 1 to 3 presented the typical types of urban agriculture in Taiwan.

Figure 1. flat agriculture

Figure 2. roof agriculture

Figure 3. vertical agriculture

Most public sector building roof has green facilities. However, the popularization rate of urban agriculture in Taiwan is still not significantly improved. The main reason is that the public cannot apply the concept of economic added value on green facilities of urban agriculture lead to ineffective result for private sectors in Taiwan. Recently, the reduction of green facilities costs created opportunities for private sectors to participate the act due to setting materials innovation. The setting of green facilities materials of urban agriculture are composed of drainage board, planting root control board, wood, and non-woven fabric shown in Figure 4.

Figure 4. Common mold materials

In order to encourage the public to use the unused space in community for farming, Taichung City government has been promoting "City Food Forest" act since 2016 to increase food security and contribute to the greening of the city as well as raise awareness of the potential of sustainability projects among citizens. "City Food Forest" act successfully implemented several cases in elementary schools those not only produced crops but also achieved the educational goals as shown in Figure 5-7.

Figure 5. Huilai elementary school demonstration

Figure 6. Four-plows elementary school demonstration

Figure7. Quiaoxiao elementary school demonstration

4 Method

Economic value added is a financial performance metric based on residual wealth, calculated by deducting a firm's cost of capital from operating profit. In essence, it is the value generated from funds invested in a business. EVA asserts that businesses should create returns during a certain period at a rate above their cost of capital. This study explored the net operating profits of both quantitative and qualitative benefits. EVA presents the following formula for its calculation:

The formula for EVA.

EVA = NOPAT - (Capital Invested x WACC)

Where NOPAT = Net Operating Profits After Tax

Capital invested = Equity + long-term debt at the beginning of the period

WACC = Weighted Average Cost of Capital

The study identified costs for green facilities of urban agriculture for individual producer that reflected the costs that project participants incur in producing food. The setting costs of green facilities on urban agriculture included 4 parts: 1.operating costs: land rent and labour, 2. tillage apparatus: wooden boxes, planting pots, support frame, 3. cultivation medium: soil, and 4. composting facilities: kitchen compost bin, as shown in table 1.

Table 1. Setting costs of green facilities

Categories	Items	Net Price
Operating costs	Land rent	No additional cost
	Labour	No additional cost
Tillage apparatus	Wooden boxes	NT$3,000 / m2
	Planting pots	NT$720 / m2
	Support frame	NT$560 / m2
Cultivation medium	Soil	NT$1,250 / m2
Composting facilities	Kitchen compost Bin	NT$300 / pcs

5 Conclusion

5.1 Urban Agriculture Benefits

The study reviewed the extant literatures and organized two quantitative factors, agricultural production and energy saving and two qualitative factors, indirect economic benefits and social and psychological benefits for the economic value added of green facilities on urban agriculture. Urban Farms may be limited by the city's design, so it may not have the formal using right of land. However, Urban Farms can plan to reuse the Green-roof to achieve the facilities green and reach its EAV. For example, this model can also be used to review the benefits of urban agriculture construction and review the probabilities of sustainable development. This study explored the practical cases to verify the interaction between qualitative and quantitative factors of green facilities contribute to urban agriculture summarized in figure 8.

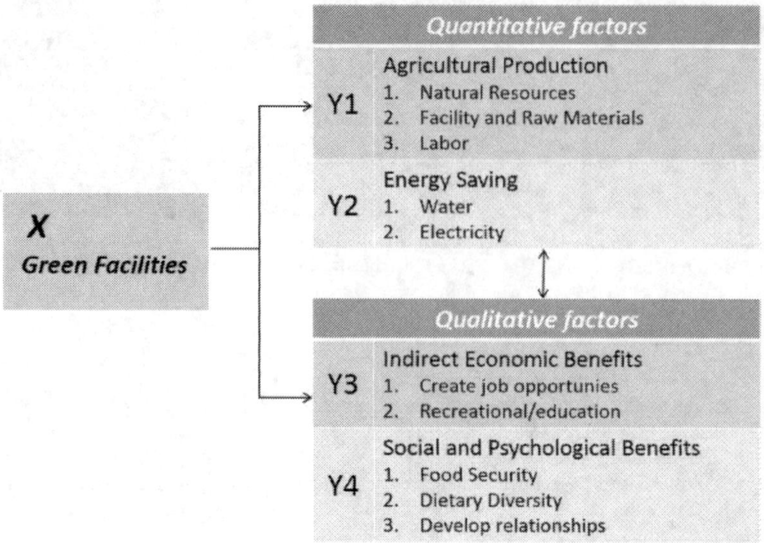

Figure 8. Interaction between qualitative factors and quantitative factors of urban agriculture impacted by green facilities.

Y1 to Y3 and Y4: Labor is one of the major costs of agricultural production. Green facilities farming might create social relationship and job opportunities.

Y3 to Y2: Green facilities of urban agriculture can provide urban farmers the training about business management, technical skills and marketing which might increase the agricultural production or save the energy by rainwater recovery system, and roof green energy.

Y4 to Y3: Urban agriculture is a marketing concept of local production and consumption which can directly identify food security and offer food nutrition and safety education.

5.2 Discussions

The most important issue for public sector when promoting green facilities of urban agriculture is not the roof planting technology but the communication with citizen for its costs and benefits. The study employed EVA to indicate the interaction between qualitative and quantitative factors that effectively offer the communication platform between public and private sectors that promoted the green technology application for urban agriculture development.

References

1. Hodgson, K. & Campbell, M.C. & Bailey, M. (2011). Urban Agriculture: Growing Healthy, Sustainable Places.
2. Health Council of the Netherlands and Dutch Advisory Council for research on Spatial Planning (2004). Nature and health. The influence of nature on social, psychological and physical well-being.
3. Qiu Yinghao. (2015). Urban Agriculture Creates New Vision of Pastoral City.
4. Kristin L. Getter &D. Bradley Rowe. (2006). The Role of Extensive Green Roofs in Sustainable Development.

The Effects of Youth Entrepreneurship Community Engagement on Sustainable Design: A Case Study of Guangfu Village Redevelopment

Hsu, *Kuo-Wei*[1], Chao, *Chieh-Yu*[2]

[1] Professor, Department of Landscape and Urban Design, Chaoyang University of Technology, No. 168, Jifeng E. Rd, Wufeng District, Taichung, 41349, Taiwan, kwhsu@cyut.edu.tw
[2] Department of Landscape and Urban Design, Chaoyang University of Technology, No 168, Jifeng E. Rd, Wufeng District, Taichung, 41349, Taiwan, yu8342@gmail.com

Abstract. Public participation plays a decisive role in the way to achieve sustainability deign for local redevelopment. Guangfu Village, the first garden city model town in Taiwan is facing redevelopment issue when executing sustainable design with human resources after 921 earthquake in 1999. Recently, youth entrepreneurship program imported 70 non-residential participants for the purpose of promoting Guangfu Village as an ecological urbanism demonstration base on garden city planning in Taiwan. Previous studies approved sustainable development was benefited with place attachment of residents. Relative literatures indicated that imported creative community brought positive effects for local redevelopment. This study focused on exploring how community engagement as a trend of human resource influenced local redevelopment when facing the challenge of sustainable design. 70 questionnaires of youth entrepreneurship community members were analysed to verify place attachment of imported community in Guangfu Village. The study furthermore applied semi-structured questionnaire to explore the effects on sustainable design for local redevelopment. According to the research analysis, community involvement factors including involvement, participation and control, local dependency factors including local dependence and local identity are the major factors for sustainable design with social, economic and ecological aspects.

Keywords. youth entrepreneurship, community engagement, place attachment, sustainable design

1 Introduction

Sustainable design practice is in need of social sustainability as well as environmental sustainability. Therefore, public participation plays a decisive role in the way to achieve sustainability deign for local redevelopment based on the identification and use of the resources and endogenous potentialities of a community. Located in cultural town, Wufeng, Guangfu Village is the first "new village" built after WWII. As the first garden city model town in Taiwan, Guangfu Village holds a special place in Taiwan's urban development history for its unique design aimed to integrate work and residential areas in green living space with the very first underground sewage system that separates rainwater and wastewater. Wufeng is full of cultural and educational atmosphere and rich in cultural resources and natural landscape, yet was heavily damaged by the 921earthquake in 1999. After the residents moved out, Guangfu Village left empty and poignant space. To drive community revitalization, the major redevelopment issue is importing sustainable design as well as human resources. Taichung City Government launched the "Taichung Youth Entrepreneurship" program and turned it into a cultural and creative village where artists and entrepreneurs came to establish their new community within the mottled historic red brick walls. Guangfu Village provided 39 buildings for 70 units as the main base of youth start-ups community. Resident participation has been proven to be an effective factor for local redevelopment. However, there still exists contradiction for residents to support green architectures when it costs more to execute sustainable design without government subsidy. On the contrary, new communities declare supporting ecological community development when they joined the program. The extent of engagement and place attachment of 70 youth entrepreneurship community members in Guangfu Village should be the key for sustainable development. The objectives of this study are as follows:

• To explore the interaction between youth entrepreneurship community participation and place attachment.
• To construct a conceptual framework of community engagement affects sustainable design.
• To examine the effects of youth entrepreneurship community engagement activities on sustainable development of Guangfu Village.

2 Research Context

The context setting of this research is to explore effects of youth entrepreneurship community engagement activities composed with interaction between community participation and place attachment on sustainable design of local redevelopment as shown in figure 1.

Figure 1 Research Context Setting

2.1. Youth Entrepreneurship Community Participation

Youth entrepreneurship community is improving employment, entrepreneurship, and civic engagement opportunities for young people to bring together their creativity and stimulate the entrepreneurial energy. Youth Entrepreneurship programs give youth access to the right resources which can effectively create opportunities for advancement and growth in labour markets while simultaneously addressing development needs. Youth entrepreneurship community participation illustrates the benefits of stimulating young people and providing them with the tools, resources or environment to put their creativities into effect. [1] [2]

2.2. Place Attachment

Place attachment consists place dependence and place identity two parts (Williams et al., 1992). [3] Place dependence arises from a positive evaluation of a place that it meets user's needs and allows them to achieve their goals; place identity concerns the meaning and significance of places for people, and how these meanings contribute to individuals' conceptualizations of self as shown in figure 2. [4] [5] [6]

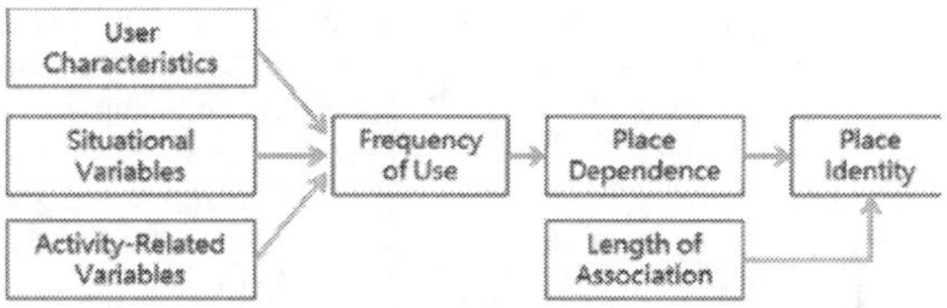

Figure 2 Place Attachment Structure

2.3. Community Engagement Activities

A community is a group of people that come together that have a common set of goals, values, and work together to achieve those common goals. Bagnall (1989) described community engagement activities for three different levels: presence, involvement and control. [7] [8] [9]

2.4. Sustainable Design

McLennan (2004) described sustainable design is the philosophy of designing physical objects, the built environment, and services to comply with the principles of social, economic, and ecological sustainability. The performance of sustainable design requires renewable resources, minimal impact on the environment, and links people to the natural environment to meet economic, social as well as ecological development. [10]

3 Research Object: From Garden City to Eco-Urbanism Demonstration

Guangfu Village, the first "new village" built after WWII was erected as a residential area for officials working in the provincial government. As the first garden city model town in Taiwan, Guangfu Village holds a special place in Taiwan's urban development history for its unique design aimed to integrate work and residential areas in green living space with the very first underground sewage system that separates rainwater and wastewater. Guangfu Village is facing redevelopment issue when executing sustainable design with human resources after 921 earthquake, one of the most serious earthquakes in Taiwan. Recently, in 1999. Youth entrepreneurship program imported 70 non-residential participants for the purpose of promoting Guangfu Village as an ecological urbanism demonstration base on garden city planning in Taiwan. Guangfu Village has turned into a cultural and creative village where young artists and entrepreneurs came to establish their creative base within the mottled historic red brick walls as shown in figure 3 and figure 4.

Figure 3 Green Resource of Guangfu Village

https://doi.org/10.1051/e3sconf/20185705002

Figure 4 Garden City Planning of Guangfu Village

4 Research Method

Both questionnaire and interview survey were used to detect the cognitive and the results presented in this study. Questionnaire design the results of the literature review serve as the basis for designing the structured questionnaire. According to the preliminary results of pre-test to amend and establish a formal questionnaire. A semi-structured questionnaire was used in this study to interview youth entrepreneurship community members of Guangfu Village to examine the interaction between community participation and place attachment for sustainable design of local redevelopment. Intensive individual interviews with youth entrepreneurship community members were made to explore their perspectives of community engagement factors and effects on the sustainable design.

4.1. Conceptual Framework

From the perspective of place attachment and community engagement, this paper explores the attitude and relevance of the youth community to the base and makes a positive analysis of the current communities in Guangfu village that have not withdrawn from the project as shown in figure 5.

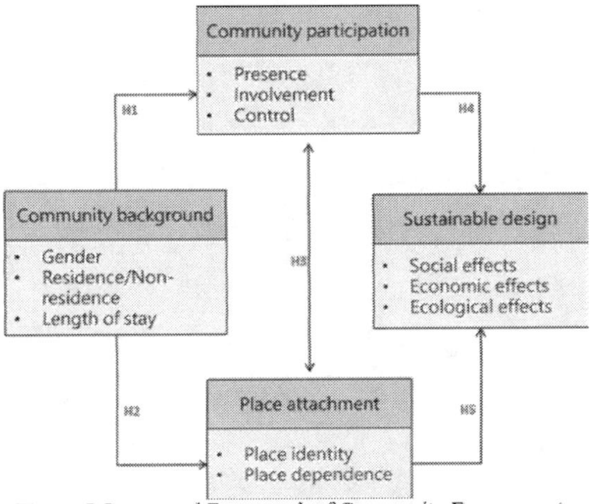

Figure 5 Conceptual Framework of Community Engagement affects Sustainable Design

4.2. Community Engagement Activities

Community engagement activities are composed of community participation and place attachment.

4.2.1 Community Participation Factors

• Presence：Means the participation of the community less involved in the activities of the region, only to participate in activities。
• Involvement ： Means participation in regional activities in the middle, including personal and social involvement in two parts. Individual means participating in activities to input opinions or opinions; society means to sharing activities and behaviors.
• Control：Means individuals with high participation in community activities, can plan or lead activists.

4.2.2 Place Attachment Factors

• Place Attachment take Williams et al. (1992) proposed a "place-dependent" and "local identity" two facets. [3]
• Place dependence：Residents' perceptions, beliefs and expectations arising from regional functional dependency.
• Place identity：Resident psychologically dependent on the feelings of the region, ideas and expectations.

5 Conclusion Research

5.1. Results

The study applied a structured questionnaire to analyst the research hypotheses and verify the youth entrepreneurship community existing strong engagement in place dependence (4.14) and medium engagement in community participation (3.38) and place identity (3.92). The community expressed strong willing to support sustainable design of local development (4.44) as shown in figure 6. A semi-structured questionnaire and interviews examined as follows: 1. No significant difference between community backgrounds affecting community engagement. 2. Resident showed more place dependence than Non-resident. 3. The extent of community participation reflected the status of place attachment. 4. The extent of community engagement activities, both of community participation and place attachment are statistical significance on the effects of sustainable design for local redevelopment.

	min	max	mean	std
Community Participation	2.63	4.00	3.3750	.54199
Place dependence	3.43	4.86	4.1429	.46070
Place identity	3.14	5.00	3.9286	.66853
Sustainable Design	4.00	5.00	4.4444	.41722

Figure 6 Descriptive Statistics of Community Engagement

 https://doi.org/10.1051/e3sconf/20185705002

5.2. Discussion

Social communities and social entrepreneurs development play an important role in local development with public sector's promotion. Social communities as well as residents support sustainable design of local development. The major issue is how to establish a platform for the interactions and dialogue between social community and residents to share the same goal for achieving local sustainable development. It is proposed that youth entrepreneurship program could establish a rolling mechanism to combine with nonprofit organization to establish an interactive mechanism in local redevelopment activities. It should recognize and attach importance to the public and the community's efforts and strengths on sustainable development.

References

1. Yusuf, J. (2012) 'A Tale of Two Exits: Nascent Entrepreneur Learning Activities and Disengagement from Start-Up', Small Business Economics 39.3: 783–99
2. Waters, L.; McCabe, M.; Kiellerup, D. and Kiellerup, S. (2002) 'The Role of Formal Mentoring on Business Success and Self-Esteem in Participants of a New Business Start-Up Program', Journal of Business and Psychology 17.1: 107–21
3. Williams, D. R., Patterson, M.E., & Roggenbuck, J.W. and Watson, A. E. (1992). Beyond the commodity metaphor: Examining emotional and symbolic attachment to place. Leisure Sciences, 14: 29-46.
4. Williams, D. R. & Roggenbuck, J.W. (1989). Measuring place attachment: some preliminary results, Paper presented at Symposium on Outdoor Recreation Plaining and Management, NRPA Symposium on Leisure Research, San Antonio, TX.
5. Moore, R. L. and Graefe, A. R. (1994). Attachments to recreation settings: the case of rail-trail users, Leisure Sciences, 16:17-31.
6. Hummon, D. M. (1992) Community attachment: local sentiment and sense of place. In: Altman, I. and Low, S. M. (eds.) Place Attachment, New York: Plenum Press, 253-278.
7. Stokols, D., Shumaker, S. A. People and Places: A Transactional View of Settings in: Harvey Cognition, Social Behavior and the Environment, 1981. pp. 441-488
8. Bagnall, R. G. (1989), Researching participation in adult education: A case of quantified distortion. International Journal of Lifelong Education, 8(3), 251-260.
9. Gottlieb, Hildy, (2007). Community Engagement Step-by-Step Action Kit, 19-20.
10. McLennan, J. F. (2004). The Philosophy of Sustainable Design.

AUTHOR INDEX